# Statistics in Social Science and Agricultural Research

## *About the Editors*

**Debasis Bhattacharya** is a Ph.D. in Statistics from the University of Calcutta. He is actively engaged in teaching and research for more than two decades. He is a Visiting Professor of the Department of Statistics in the University of California at Davis, USA. He has to his credit a significant number of research papers published in national and international journals of repute. He authored several books in Statistics, Probability and Statistical Inference. Dr. Bhattacharya visited different universities which include the Korea University at Seoul, South Korea, National University of Singapore, Singapore, Fu-Jen University, National Cheng-Kung University at Taiwan, Kazakhstan Institute of Management, Economics and Strategic Research, Republic of Kazakhstan, Stanford University, Rice University, Purdue University, USA. Dr. Bhattacharya is the Managing Editor of International Journal of Statistics and Managment Systems. At present he is a Professor in Visva-Bharati University, Santiniketan, India.

**Soma Roychowdhury** did her Ph.D. from the Department of Statistics, University of Calcutta. She is the author of several books in Statistics, Probability and Statistical Inference and research papers published in reputed national and international journals. She is the recipient of Gold centered medal for standing first among the female candidates in the Secondary Examination of West Bengal Board of Secondary Education. Dr. Roychowdhury is a Visiting Professor of the Department of Statistics, University of California at Davis, USA. She visited Stanford University, USA, Korea University at Seoul, National University of Singapore, Fu-Jen University, National Cheng-Kung University, Taiwan. At present she is Head of the Department of Public Systems Management, Indian Institute of Social Welfare and Business Management, Calcutta, India.

# Statistics in Social Science and Agricultural Research

*Edited by*
**Debasis Bhattacharya**
*Visva-Bharati University*
*Santiniketan, India*

and

**Soma Roychowdhury**
*Indian Institute of Social Welfare and*
*Business Management, Kolkata, India*

CONCEPT PUBLISHING COMPANY PVT. LTD.
NEW DELHI-110059

**ISBN-13 : 978-81-8069-822-4**

First Published 2012

*Published and Printed by*

Concept Since 1974

**Concept Publishing Company Pvt. Ltd.**
A/15-16, Commercial Block, Mohan Garden
New Delhi-110059 (India)
*T*: 25351460, 25351794, *F*: 091-11-25357109
E: publishing@conceptpub.com, W: www.conceptpub.com
**Editorial Office:** H-13, Bali Nagar, New Delhi-110 015, India.

Cataloging in Publication Data--*Courtesy:* D.K. Agencies (P) Ltd. <docinfo@dkagencies.com>

Statistics in social science and agricultural research / edited by Debasis Bhattacharya and Soma Roychowdhury.
p. cm.
Includes bibliographical references and index.
ISBN 9788180698224

1. Social sciences—Statistical methods—Congresses. 2. Social sciences—Research—Methodology—Congresses. 3. Agriculture—Statistical methods—Congresses. 4. Agriculture—Research—Methodology—Congresses. I. Bhattacharya, Debasis, 1967- II. Roychowdhury, Soma.

DDC 519.5 23

# Preface

This book discusses the uses of various statistical tools and techniques used in the research areas of social science, agriculture, environmental science, medical science and others. Any empirical research is incomplete without proper inference; and which heavily depends on the use of appropriate statistical techniques. Thus the role of statistical methods in theory and practice is very important. This volume puts together a set of scientific papers presented by many leading statisticians of national and international repute (from India and abroad) in the IMS-sponsored mini-meeting-cum-workshop held at Visva-Bharati University, Santiniketan. Participants from different fields of Statistics, Agricultural Science and Social Science including the field of medical research discussed various issues arising out of the context of social science and agricultural science and presented the tools for resolving the issues. Empirical studies and theoretical developments both have been suitably blended in their papers. Efforts have been made to keep two distinct divisions, one containing the papers related to social science issues and the others are related to issues of importance from the field of agricultural science. Inherent complexity in the data arising out of the context of social science and agricultural science research makes it difficult to apply standard statistical tools directly. Variability in the data exists at different levels because of the dependence, of unknown nature, qualitative nature of the data, non-formal distribution of the underlying variables; errors in the measurements, unavailability of some observations etc. are some of the many reasons which prevent the direct application of the existing statistical techniques.

The papers addressed some of the issues stated above and will give a very good guidance to the researchers applying statistical techniques in analyzing data of different fields of study. No doubt, this volume will be very useful to the researchers and postgraduate students in the field of social science, agricultural science and statistics.

The mini-meeting-cum-workshop could not have taken place without generous financial support from the Institute of Mathematical Statistics (IMS). The Institute of Mathematical Statistics is an International Society which fosters the development and dissemination of the theory and applications of statistics and probability. We express our sincere gratitude to IMS for sponsoring this event at Santiniketan. We are thankful to all the contributors for their valuable contributions without which this volume would not have been possible to publish. We hope, it would be found useful to all those scientists and practitioners engaged in the fields of social science, agricultural science and others. Although we have made a good deal of efforts in reviewing the articles for improving the quality, both in terms of correctness and presentation, but responsibility of the views expressed and conclusions made in the papers lies on the respective authors.

Finally, thanks go to the Concept Publishing Company Pvt. Ltd. for bringing out this volume for furtherance of scientific research in the areas of social science and agricultural sciences.

Santiniketan, India — *Editors*
Kolkata, India — Debasis Bhattacharya
June, 2011 — Soma Roychowdhury

## Contents

# 1

# On a Comparison of Randomized Response Technique and Group Response Technique for Eliciting Sensitive Information

*Yogendra P. Chaubey*

## 1. Introduction

When we are dealing with surveys containing sensitive or personal questions (e.g. drug use, alcohol abuse, abortion, illegitimate birth, contact with organized crime, sexual activity and aberrations, child abuse, mental illness, shoplifting and violation of income tax law, etc.) as quality attributes we are generally faced with a significant proportion of non-response and/or responses, which may not be true. Warner (1965) developed the Randomized Response Technique (RRT) for estimating the proportion of "positive" responses in such a situation. This technique uses a randomized device and asks the respondent to answer the sensitive question only if the outcome of the randomized device is the one specified in advance. Thus, this technique protects the privacy of the respondents without releasing their personal status with respect to the sensitive question. Hence, non-response rate may be expected to be significantly reduced, and respondents may not be fearful or embarrassed of their status with respect to the sensitive question.

However, this technique may have some limitations and for this reason several modifications have been proposed in the literature (Greenberg et al., 1986). Chaubey and Li (1995b) put forward an alternative technique called the Group Response Technique (GRT). This technique involves taking samples in batches, and classifying them into one of two mutually exclusive categories, such as positive and negative or defective and non-defective. Each batch is supposed to consist of an equal number of sampling units and may be natural or artificially constructed. A composite sample of size $k$ is formed by physically mixing $k$ individual test portions (units), then a single measurement is made declaring the composite to be positive or negative. Alternatively, a sample of size $n$ may be randomly divided into $m$ batches, each consisting of $k$ units such that $n = mk$. Thus, in batch sampling, a batch is declared positive if at least one of its members has positive characteristics. Traditionally, each unit of a sample is classified and therefore, batch sampling is more cost effective compared to the traditional approach.

Because the individual responses to a sensitive question are not recorded in GRT, total anonymity of the respondents in GRT is guaranteed as in composite sampling (Hammick and Gastwirth, 1994). Hence GRT presents itself as a valid alternative to RRT.

Garner et al. (1987) considered the maximum likelihood estimation of the population prevalence using batch sampling and advocated its use in environmental monitoring. Chaubey and Li (1995a) further studied the maximum likelihood estimator with respect to its bias and proposed a modification. They also investigated the Bayesian method of estimation when some *apriori* information about the unknown proportion is available. The reader may further be referred to the papers by Boswell and Patil (1987) and Boswell et al. (1992) for use of the above technique in classification of polluted and non-polluted samples. Chaubey et al. (1997) proposed this method as a tool in estimating quality, which is interpreted as proportion, such as fraction defective or proportion of dissatisfied customers.

In this paper we review the comparative advantages of RRT and GRT and present an empirical study conducted in

China and Concordia University. The maximum likelihood estimator and its bias correction have been presented in Section 2, and Section 3 presents a description of Warner's original design along with the Bayesian estimation due to Franklin and Winkler (1979). In Section 4, we present the Bayes' estimators as given in Chaubey and Li (1995a) for GRT. Section 5 compares the mean squared errors of different estimators for several sets of parameter values along with a comparison of Bayes' risk for different Bayes' estimators as reported in Chaubey et al. (1997). Finally, Section 6 presents a reliability and validity study of these estimators in the context of an empirical study carried out in Li (1995).

## 2. Randomized Response Technique

### 2.1 Warner's Design

In the original RRT (known as Warner's design), a group of respondents are asked the sensitive question and are instructed to answer it only if the outcome of the randomizing device is "positive". Let $P$ denote the probability of "positive" outcome of the randomizing device and $r$ individuals are recorded to be having the attribute of interest then $\hat{\lambda} = r/n$ estimates the quantity

$$\lambda = \pi_A P + (1 - \pi_A)(1 - P) \qquad \dots (2.1)$$

unbiasedly. Thus, an unbiased estimate of $\pi_A$ using Warner's design is given by

$$\hat{\pi}_{AW} = \frac{(r/n) - (1 - P)}{2P - 1} \,. \qquad \dots (2.2)$$

Clearly, $P = 1$ amounts to direct response and $P = ½$ renders the technique useless. The variance of the above estimator to be used for comparison is given by

$$V(\hat{\pi}_{AW}) = \frac{\pi_A(1-\pi_A)}{n} + \frac{P(1-P)}{n\,(1-2P)^2}. \qquad \dots (2.3)$$

### 2.2 Unrelated Question Design

A simple modification of Warner's design is due to Simmons {see Greenberg et al. (1969)}, commonly known as the Unrelated Question Design in which the respondents are asked to answer if they belong to another group *U* (unrelated to the group of the basic attribute) in case the outcome of the randomizing device is not positive. Letting $\pi_U$ denote the proportion of individuals belonging to the group *U*, *r/n* estimates the quantity

$$\lambda_U = P\pi_A + (1-P)\,\pi_U.$$

Assuming that $\pi_U$ is known, an unbiased estimator of $\pi_A$ is given by

$$\hat{\pi}_A = \frac{(r/n)-(1-P)\pi_U}{P}. \qquad \dots (2.4)$$

The variance of the estimator for comparison purpose may be given by

$$V(\hat{\pi}_A) = \frac{\lambda_U(1-\lambda_U)}{nP^2}. \qquad \dots (2.5)$$

### 3. Group Response Technique

Suppose random samples of size *k*, i.e., groups of size *k*, are grouped from the *n* individuals in the sample ($n >> k$), and each group will use a group answering device to answer questions of the survey at the same time. Let $\pi_A$ denote the proportion of individuals in the population with the specific characteristic (belonging to group A). Let $m = n/k$ and $x_{(k)}$ be the number of groups which possess the characteristic, then $x_{(k)} \sim$ Binomial $(m, \pi'_A)$, where

$$\pi_A = 1 - (1 - \pi_A)^k \qquad \text{... (3.1)}$$

and therefore,

$$\pi_A = 1 - (1 - \pi'_A)^{1/k} \qquad \text{... (3.2)}$$

### 3.1 Maximum Likelihood Estimator of $\pi_A$ in GRT

The MLE of $\pi'_A$ as in Garner et al. (1989) is given by $\hat{\pi}'_{Ak} = X_k/m$ and thus the MLE of $\pi_A$ is given by

$$\hat{\pi}_{Ak} = 1 - (1 - X_k/m)^{1/k}. \qquad \text{... (3.3)}$$

It should be noted that the MSE of the MLE can be minimized with respect to *k*. These are summarized in Table 1.1 (See Garner et al., 1969). Thus the optimum value of *k* depends on the unknown value of $\pi_A$. Hence the value of *k* should be chosen depending on some prior guessed value of $\pi_A$.

**Table 1.1** Optimal *k* for given $\pi_A$ and *m*

| | *m* | | | | | | |
|---|---|---|---|---|---|---|---|
| $\pi_A$ | 10 | 20 | 30 | 40 | 50 | 70 | 100 |
| 0.25 | 3 | 3 | 4 | 4 | 5 | 5 | 5 |
| 0.10 | 5 | 8 | 10 | 12 | 12 | 13 | 13 |
| 0.05 | 9 | 15 | 20 | 25 | 25 | 25 | 25 |
| 0.01 | 35 | 60 | 70 | 110 | 120 | 140 | 140 |

The estimator in (3.3) is not unbiased, unlike the case for RRT. The bias correction as given in Chaubey and Li (1995a) is presented below. Even though, the explicit form of bias as a function of $\pi_A$ is not available, a reasonable approximation may be provided using the standard methods. We have, for large *m*,

$$E(\hat{\pi}_{Ak}) \approx \pi_A + \frac{k-1}{2mk^2}\left[\frac{1-(1-\pi_A)^k}{(1-\pi_A)^{k-1}}\right] \quad \text{... (3.4)}$$

The comparison of approximate bias using the formula (3.4) with the exact bias shows that the above approximation is adequate for our purposes [See Table 2.2 of Chaubey and Li (1995a)]. Thus, we propose a bias corrected estimator, $\hat{\pi}_{Ak(C)}$, given by

$$\hat{\pi}_{AK(c)} = \pi_{AK} - \frac{k-1[X_{(k)}/m]}{2mk^2\left[1\frac{X_{(k)}}{m}\right]^{\frac{(k-1)}{k}}} \quad \text{... (3.5)}$$

Chaubey and Li (1995a) summarized the ratio of bias of the bias corrected MLE to that of the uncorrected MLE for various sets of values ($\pi_A$, $m$, $k$) from which it is observed that the bias correction is effective for small as well as large $k$. Substantial reduction in bias is observed for small values of $\pi_A$. Chaubey and Li (1995a) also found that the bias corrected estimator, generally, has higher efficiency as compared to the uncorrected MLE and whenever there is a loss in efficiency it is negligible.

### 3.2 Bayesian Estimation in GRT

In the following section we will consider the Bayesian estimation procedure in GRT. To develop a Bayes estimator for $\pi_A$ in GRT we may have two choices, namely, to consider a prior on $\pi_A$ or a prior on $\pi'_A$. In our opinion the prior on $\pi_A$ is more meaningful from practical considerations since $\pi'_A$ depends on $k$, whose optimal value is desired for producing the smallest MSE using sample compositing. However, when the value of $k$ is fixed and the experimenter has some knowledge of $\pi'_A$, the alternative prior may also be meaningful. For completeness, both cases are considered in this paper.

### *3.2.1 Bayesian Estimator with Prior on $\pi_A$*

Consider a Beta(α,β) prior on $\pi_A$ which can accommodate a variety of prior information about $\pi_A$ for developing Bayes estimator, i.e., we consider a probability distribution on $\pi_A$ given by

$$f(\pi_A) = \frac{1}{B(\alpha,\beta)} \pi_A^{\alpha-1} (1-\pi_A)^{\beta-1} \quad \dots (3.6)$$

The beta distributions may be viewed as reasonable approximations for a wide variety of types of prior information on $\pi_A$ (Winkler and Franklin, 1979).

In the following section, sometimes $\pi$ will be used instead of $\pi_A$, $x$ will be substituted for $x_k$ and X will be substituted for $X_k$ if there is no possibility for any confusion.

**Theorem 3.1 :** When Beta (α,β) prior is put on $\pi_A$, the Bayes estimate of $\pi_A$ is given by

$$\hat{\pi}_{Ak(B)} = \frac{\sum_{j=0}^{x} \binom{x}{j} (-1)^j B(\alpha+1, kj+km-kx+\beta)}{\sum_{j=0}^{x} \binom{x}{j} (-1)^j B(\alpha, kj+km-kx+\beta)} \quad \dots (3.7)$$

### *3.2.2 Bayesian Estimator with Prior on $\pi'_A$*

**Theorem 3.2 :** When Beta prior is put on $\pi'_A$, the Bayes estimate of $\pi_A$ is given by

$$\hat{\pi}_{Ak(B)} = 1 - \frac{\Gamma(m+\alpha+\beta)\Gamma(m-x+\beta+1/k)}{\Gamma(m-x+\beta)\Gamma(m+\alpha+\beta+1/k)} \qquad \ldots (3.8)$$

An ad hoc but simple estimation procedure would be to consider the posterior mean of $\pi'_A$ and use it to consider an estimator of $\pi_A$ by virtue of the relation between $\pi_A$ and $\pi'_A$. This procedure will be called indirect Bayesian procedure and is presented in the next section.

### 3.2.3 Indirect Bayes Estimator $\hat{\pi}'_{Ak(IB)}$

Choosing a $B(\alpha, \beta)$ prior on $\pi'_A$ gives the Bayes estimator of $\pi'_A$ as

$$\hat{\pi}'_A = \frac{X_{(k)} + \alpha}{m + \alpha + \beta} \,. \qquad \ldots (3.9)$$

Substituting it into (3.2) in place of $\pi'_A$, we have indirect Bayesian (IB) estimator :

$$\hat{\pi}_{Ak(IB)} = 1 - (1 - \frac{X_{(k)} + \alpha}{m + \alpha + \beta})^{1/k} \,. \qquad \ldots (3.10)$$

The estimator in (3.10) can be given the following justification. For large $N$ we have from Abramowitz and Stegun (1964, Eq. 6.1.46)

$$\frac{\Gamma(N+a)}{\Gamma(N+b)} \approx N^{a-b} \,. \qquad \ldots (3.11)$$

Therefore, for large $m$, from (3.9) we find that

$$\hat{\pi}_{Ak(B)'} \approx 1-(\frac{m-X_{(k)}+\beta}{m+\alpha+\beta})^{1/k} = \hat{\pi}_{Ak(IB)} \qquad \ldots (3.12)$$

$\pi_{AK(IB)}$ is easier to compute, and therefore may be preferred in practice.

In the case of Uniform prior, which is a special case of the above, (3.10) simplifies to

$$\hat{\pi}_{4k(IB)U} = 1-(1-\frac{X_k+1}{m+2})^{1/k}, \qquad \ldots (3.13)$$

which will be considered for comparison purpose, where subscript $U$ means uniform prior. In Section 4, we present some comparisons of the estimators mentioned above.

## 4. Comparison of Estimators

### *4.1 Estimators in GRT*

#### *Bayes Risk Comparison*

In general, the Bayes estimator $\delta_B(X)$ is an "optimal" estimator of the parameter $\theta$ in the sense that under the given prior and loss function it has the smallest risk, where the risk $r(\tau,\delta)$ for a decision rule (estimator) $\delta(X)$ with respect to the prior under the squared error loss $\tau$ is given by

$$r(\tau,\delta)=E_\theta E_{X\theta}[\delta(X)-\theta]^2 \qquad \ldots (4.1)$$

Therefore, the task of a Bayesian is done once the Bayes estimator is computed. However, it may be of interest to see how far are the other estimators which may have frequentist's appeal in terms of the Bayes risk or how does the Bayes estimator fare with respect to the standards of a frequentist. For this purpose, first we compare the Bayes risks of $\hat{\pi}_{Ak(B)}$

and $\hat{\pi}_{Ak(B)'}$ with that of the MLE. We define the Bayes Relative Efficiency (BRE) of an estimator $\delta$ by

$$BRE(\delta) = \frac{r(\tau,\delta_B)}{r(\tau,\delta)} \qquad \ldots (4.2)$$

For comparing the Bayes estimator corresponding to a Beta($\alpha,\beta$) prior on $\pi_A$ the prior on $\pi'_A$ is chosen to be Beta($\alpha',\beta'$), where $\alpha'$ and $\beta'$ are obtained by equating the first two moments. Thus we have

$$\alpha' = \frac{(1-A)(B-A)}{A^2-B}, \beta' = \frac{A(B-A)}{A^2-B}, \qquad \ldots (4.3)$$

where $A = \Gamma(\alpha+\beta)\Gamma(\beta+k)/\Gamma(\beta)\Gamma(\alpha+\beta+k)$

and $B = \Gamma(\alpha+\beta)\Gamma(\beta+2k)/\Gamma(\beta)\Gamma(\alpha+\beta+2k)$.

We summarize below the conclusions regarding these comparisons from Chaubey and Li (1995a).

- The MLE and the bias corrected MLE are much inferior to the Bayes estimators, using equivalent priors on $\pi'_A$ as on $\pi_A$.
- The Bayes estimator with matching prior on $\pi'_A$ with that on $\pi_A$ gives almost matching risk.
- The three alternative estimators have 100% BRE for k=1, because these coincide with the Bayes estimator.
- The ad hoc estimator seems to be a good approximation to $\hat{\pi}_{Ak(B)'}$ especially when $\pi_A$ is small.

*Classical Comparison*

We may also provide a frequentist's assessment of various estimators incorporating prior information with respect to. MLE. The criteria used for this purpose are relative bias (RB) and relative efficiency (RE), which are defined as follows. The relative bias of an estimator $\hat{\pi}_A$ is defined to be

$$RB(\hat{\pi}_A) = \frac{E(\hat{\pi}_A - \pi_A)}{\pi_A}, \qquad \text{... (4.4)}$$

whereas the relative efficiency of $\hat{\pi}_A$ is given by

$$RE(\hat{\pi}_A) = \frac{MSE(\hat{\pi}_{Ak})}{MSE(\hat{\pi}_A)}. \qquad \text{... (4.5)}$$

These may be termed as frequentist comparisons. In these comparisons, the Bayes method is used only to get the form of the estimator, and loss function does not play any further role. As mentioned before, we want to assess the performance of different estimators by frequentist measures.

- The biases of the estimators $\hat{\pi}_{AK(B)}$, $\hat{\pi}_{AK(B')}$ and $\hat{\pi}_{AK(IB)}$ are smaller than those of the MLE and its modified version.
- Indirect Bayes estimator performs very well in the sense of having small bias.
- The biases of both the Bayes estimators with matching priors almost coincide.
- MSE's of Bayes estimators, $\hat{\pi}_{AK(B)}$, $\hat{\pi}_{AK(B')}$ and $\hat{\pi}_{AK(IB)}$ are smaller than those of MLE and its corrected version.
- The Indirect Bayes estimator performs very well in the sense of having small MSE; it gives MSE even smaller than the bias corrected MLE.

### *4.2 Comparison Among Estimators in GRT and RRT*

#### *4.2.1 Comparison among the MLE's*

Chaubey and Li (1995b) compared the MLE in GRT with that in RRT using Warner's and Simmon's designs under various configurations of parameters based on the mean squared errors of the estimators. They derive the following conclusions :

- GRT has higher efficiency than RRT using Warner's design for $P = .6, .7, .8$ and $\pi_A < .25$.
- The relative efficiency of GRT with respect to RRT using Warner's design increases as the value of $\pi_A$ decreases.
- Substantial gains in relative efficiency using GRT over RRT under Warner's design are observed for small values of $\pi_A$.
- GRT gives higher efficiency over RRT using Simmon's design for $P = .6$ and $\pi_A < .25$, $P = .7$ and $\pi_A < .05$, $P = .8$ and $\pi_A < .01$.
- Noticeable gains in relative efficiency using GRT over RRT under Simmon's design are observed for small values of $\pi_A$.

### *4.2.2 Comparison among the Bayesian estimators in GRT and RRT*

In this section we assess the relative performance of Bayesian estimators in GRT and RRT according to the criteria of relative efficiency. Here we have considered only Warner's design in RRT. The Bayes estimator using the prior on $\pi'_A$ is also not considered because it is observed in Section 4.1.1 to be risk equivalent to the Bayes estimator under induced prior on $\pi'_A$.

We use a Beta prior on $\pi_A$ and consider the approximate Bayes estimator under Warner's design in RRT and indirect Bayes estimator in GRT using this prior. Table 1.2 presents the MSE's of Bayes estimator under various priors and sample sizes for GRT and RRT. The value of $k$ is chosen to be near optimal for the use of the Bayes estimator in GRT. Substantial reduction in MSE is noted, especially for small values of $\pi_A$ for which the priors considered here may be appropriate. We provide a similar comparison for MLE and the Bayes estimator using RRT in Table 1.3, where the results are provided for uniform prior only.

**Table 1.2 :** MSE's of Bayes Estimators in RRT and GRT for $P = 0.7$

| $(\alpha,\beta)$ | | (1,1) | | (1,10) | | (1,100) | | (1,1000) | |
|---|---|---|---|---|---|---|---|---|---|
| $n$ | $\pi_A$ | $\hat{\pi}_{A(WB)}$ | $\hat{\pi}_{Ak(B)}$ | $\hat{\pi}_{A(WB)}$ | $\hat{\pi}_{Ak(B)}$ | $\hat{\pi}_{A(WB)}$ | $\hat{\pi}_{Ak(B)}$ | $\hat{\pi}_{A(WB)}$ | $\hat{\pi}_{Ak(B)}$ |
| 150 | 0.25 | .0087 | .0021 | .0102 | .0084 | .0564 | .0443 | .0620 | .0602 |
| | 0.10 | .0092 | .0007 | .0023 | .0003 | .0082 | .0062 | .0098 | .0095 |
| | 0.05 | .0089 | .0004 | .0026 | .0002 | .0018 | .0014 | .0024 | .0024 |
| | 0.01 | .0086 | .0001 | .0036 | .00004 | .00004 | .00004 | .00009 | .00009 |
| | 0.001 | .0086 | .00005 | .0039 | .00003 | .00005 | 9.4E-7 | 4.6E-7 | 6.1E-7 |
| 300 | 0.25 | .0047 | .0011 | .0059 | .0036 | .0552 | .0337 | .0620 | .0581 |
| | 0.10 | .0046 | .0004 | .0019 | .0004 | .0079 | .0044 | .0098 | .0091 |
| | 0.05 | .0042 | .0002 | .0020 | .0001 | .0017 | .0010 | .0024 | .0021 |
| | 0.01 | .0038 | .00004 | .0025 | .00003 | .00004 | .00003 | .00009 | .00009 |
| | 0.001 | .0038 | .00001 | .0026 | 9.7E-6 | .00005 | 8.5E-7 | 4.8E-7 | 5.8E-7 |
| 450 | 0.25 | .0032 | .0008 | .0040 | .0021 | .0541 | 0.0267 | .0620 | .0561 |
| | 0.10 | .0031 | .0002 | .0016 | .0003 | .0077 | .0033 | .0098 | .0086 |
| | 0.05 | .0028 | .0001 | .0017 | .0001 | .0016 | .0007 | .0024 | .0021 |
| | 0.01 | .0024 | .00003 | .0018 | .00002 | .00004 | .00002 | .00009 | .00008 |
| | 0.001 | .0023 | 6.9E-6 | .0019 | 5.4E-6 | .00005 | 7.7E-7 | 4.8E-7 | 5.5E-7 |

**Table 1.3** MSE's of MLE and Bayes Estimators in RRT under Uniform Prior

| $\pi_A$ | 0.25 | | 0.1 | | 0.05 | | 0.01 | |
|---|---|---|---|---|---|---|---|---|
| | $\hat{\pi}_{AW}$ | $\hat{\pi}_{A(WB)}$ | $\hat{\pi}_{AW}$ | $\hat{\pi}_{A(WB)}$ | $\hat{\pi}_{AW}$ | $\hat{\pi}_{A(WB)}$ | $\hat{\pi}_{AW}$ | $\hat{\pi}_{A(WB)}$ |
| $n$ | | | | $P = 0.6$ | | | | |
| 150 | .0413 | .0285 | .0406 | .0370 | .0403 | .0391 | .0401 | .0407 |
| 300 | .0206 | .0164 | .0203 | .0197 | .0202 | .0201 | .0200 | .0204 |
| 450 | .0138 | .0116 | .0135 | .0132 | .0134 | .0132 | .0134 | .0130 |
| $n$ | | | | $P = 0.7$ | | | | |
| 150 | .0100 | .0087 | .0094 | .0092 | .0091 | .0089 | .0088 | .0086 |
| 300 | .0050 | .0047 | .0047 | .0046 | .0045 | .0042 | .0044 | .0038 |
| 450 | .0033 | .0032 | .0031 | .0031 | .0030 | .0028 | .0029 | .0024 |
| $n$ | | | | $P = 0.8$ | | | | |
| 150 | .0042 | .0040 | .0036 | .0036 | .0033 | .0032 | .0030 | .0027 |
| 300 | .0021 | .0021 | .0018 | .0018 | .0016 | .0016 | .0015 | .0012 |
| 450 | .0014 | .0014 | .0012 | .0012 | .0011 | .0011 | .0010 | .0008 |

We find that Bayes estimator in RRT is better than MLE even when the prior is uniform. Further, from Table 1.2 we find that the Bayes estimator in RRT has lower MSE's than that in GRT under a variety of priors. Thus, we can conclude that the Bayes estimator in GRT is generally better than MLE as well as the Bayes estimator in RRT. This provides another reason for the preference of GRT over RRT. The comparison is limited to the values of $\pi_A < .25$, however, it should be noted that in many situations involving sensitive questions the proportions may not be too high.

## 5. An Empirical Study of Reliability and Validity of RRT and GRT

### *Reliability*

Reliability concerns the extent to which an experiment or test yields the same results on repeated trials. Higher consistency of the results is interpreted as higher reliability. This may be measured by a correlation coefficient between two tests measuring the same object.

### *Validity*

Validity of a test is judged by determining whether it actually measures what it proposes to measure. Out of several types of validities we consider content, criterion related and construct validity.

### *Content Validity*

It examines the content of the test directly in relation to the objective to be quantified. This is not relevant here in measuring proportion; as long as the question is clear there is no problem.

### *Criterion Validity*

This type of validity explores the relation between a related criterion to the objective such as in IQ tests.

### *Construct Validity*

This refers to the extent to which a test measures the theoretical construct or trait it was supposed to measure. This is more appropriate in the present context. This is obtained through factor analysis.

### *5.1 Study Design*

The data were collected from four schools labeled as Schools A, B, C and D. Schools A and B were comprehensive universities with approximately 10,000 students in each. School C was a technical college and School D was a high school. Table 1.4 shows the distribution of samples in different schools along with the technique used for estimating the proportions of different sensitive questions.

### *5.2 Sampling Scheme*

The sampling scheme used was a two-stage sampling, the first stage being cluster sampling, the clusters being different sections of classes and the second stage being the simple random sampling without replacement. The sample size was taken proportional to the cluster size. Table 1.4 shows the distribution of samples in different schools along with the technique used for estimating the proportions for different questions.

### *5.3 Sampling Procedure*

In School A, classes in second to fourth year in each academic

**Table 1.4 :** Distribution of samples in four schools in Shanghai

| *School* | *Dept.* | *RRT* | *GRT* | *Q1* | *Q2* | *Q3* | *Q4* | *Q5* | *Q6* | *Q7* | *Q8* |
|---|---|---|---|---|---|---|---|---|---|---|---|
| A | English | 67 | | * | * | | | | | | |
| | Politics | 54 | | * | * | | | | | | |
| | History | 40 | | * | * | | | | | | |
| | Environment | 23 | | * | * | | | | | | |
| | Statistics | 79 | | * | * | * | * | | | | |
| | Mathematics | 110 | | * | * | | | | | | |
| | Biology | 64 | | * | * | | | | | | |
| | Physics | 39 | | * | * | | | | | | |
| | Chemistry | 35 | | * | * | | | | | | |
| B | Physics | | 44 | * | * | | | | | | |
| | Mathematics | | 84 | * | * | | | | | | |
| | Chinese | | 72 | * | * | * | * | | | | |
| | Geography** | 45 | 64 | * | * | * | * | * | * | * | * |
| | English | | 44 | * | * | * | * | | | | |
| | Politics | 59 | | * | * | * | * | * | * | * | * |
| C | | 29 | | * | * | * | * | * | * | * | * |
| D | | 93 | | * | * | * | * | * | * | * | * |
| Total | | 737 | 308 | | | | | | | | |

* Inclusion in the questionnaire.

** There were 34 students who participated in both, RRT and GRT.

department were chosen randomly. The randomization device for RRT was simulated by a bag with 5 ping-pong balls, 3 labelled A and 2 labelled B. The student would answer the sensitive question if the selected ball was A, thus $P$=0.6, otherwise the student was instructed to answer the unrelated question. For some practical reasons only two questions were given to each student except in the Statistics department.

In Schools B, C and D the randomized device was simulated by throws of a die. If the rolled number was 1 or 2, the unrelated question would be answered, otherwise the sensitive question would be answered. Thus $P$ = .67. For these three schools Questions 5-8 were the same as Questions 1-4, respectively, but the unrelated questions were different.

Because of this particular design Q5-Q8 could be considered as alternative forms of Q1-Q4 in RRT.

## Sensitive Questions

Q 1. Do you believe that it is alright for students to cheat on academic tests or Exams?
*Unrelated Question :* Did you get an even number on the die?

Q 2. Did you cheat on any test or exam in the college or university?

Q 3. Do you approve of student's reading and watching pornography before the age of 18?

Q 4. Did you read or watch pornography before you were 18?

GRT was administered only in School B. A double blind procedure was used in School B. The interviewers were chosen so that they did not know any students. The students were asked to push a light switch if their answer was "yes" to the sensitive question. The interviewer would record whether the hidden light under the desk was turned on, all the switches for a group were connected to this single light, so that at lowest one response would be enough to turn the light on. The students in the group would not have information about the groups, answer.

In this study there are two types of sensitive questions; two of the four questions are about academic honour and the other two are about pornography. Each type has two questions, one for action and the other for belief.

## Reliability of RRT

Correlation between alternative forms may be used to assess the reliability of an instrument. Suppose the data are considered to be obtained under a correlated Bernoulli distribution :

$$b(x,y) = p_1^x q_1^{1-x} p_2^y q_2^{1-y}\left[1 + d\frac{(x-p_1)(y-p_2)}{p_1 q_1 p_2 q_2}\right],$$

where, $q_i = 1 - p_i$, $i = 1,2$, and $d = \rho\,(p_1 q_1 p_2 q_2)$. Let the data obtained be as follows :

| X | Y | $\bar{B}$ 0 | B 1 | Total |
|---|---|---|---|---|
| $\bar{A}$ | 0 | $n_{00}$ | $n_{01}$ | $n_{0\cdot}$ |
| A | 1 | $n_{10}$ | $n_{11}$ | $n_{01\cdot}$ |
| Total | | $n_{\cdot 0}$ | $n_{\cdot 1}$ | n |

Then a measure of reliability (Li, 1995) is given by

$$\bar{P} = \frac{n_{00}n_{11} - n_{01}n_{10}}{\sqrt{n_{0\cdot}n_{1\cdot}n_{\cdot 0}\,n_{\cdot 1}}}.$$

This is summarized for data for School D below:

| Q1&5 | Q2&6 | Q3&7 | Q4&8 |
|---|---|---|---|
| 0.98 | 0.84 | 0.63 | 0.98 |

Alternatively, a combined measures is given by the intra-class correlation coefficient, $\hat{P}_I = (MSB - MSW)/(MSB + MSW)$, where MSW and MSB are between and within mean squares, treating the estimates obtained from a one way random effects model, with four categories. The intra-class correlations for Schools B, C and D are obtained as, respectively, 0.86, 0.46 and 0.93. The small reliability coefficient for School C points to the possible large variation in the estimates due to smaller sample size.

## Split Sample Reliability

Correlations between responses of split halves samples may be used when the alternative form of test is not available. In this case it is not possible since the true responses are unknown. However, the stability of the estimates may be judged by random splitting of the data which may provide

**Table 1.5** : Estimates of $\pi_A$ in four schools in Shanghai using RRT

| *School* | *Dept.* | *No.* | *Q1* | *Q2* | *Q3* | *Q4* | *Q5* | *Q6* | *Q7* | *Q8* |
|---|---|---|---|---|---|---|---|---|---|---|
| A | English | 67 | 0.090 | 0.338 | | | | | | |
| | Politics | 54 | 0.222 | 0.160 | | | | | | |
| | History | 40 | 0.167 | 0.292 | | | | | | |
| | Environment | 23 | 0.247 | 0.827 | | | | | | |
| | Statistics | 79 | 0.093 | 0.287 | 0.308 | 0.137 | | | | |
| | Mathematics | 110 | 0.227 | 0.440 | | | | | | |
| | Biology | 64 | 0.032 | 0.500 | | | | | | |
| | Physics | 39 | 0.308 | 0.137 | | | | | | |
| | Chemistry | 35 | 0.428 | 0.333 | | | | | | |
| B | Geo & Pol. | 104 | -0.019 | 0.370 | 0.100 | 0.313 | 0.01 | 0.319 | 0.143 | 0.169 |
| C | | 29 | 0.267 | 0.733 | 0.526 | 0.785 | 0.371 | 0.888 | 0.071 | 0.446 |
| D | | 93 | 0.170 | 0.761 | 0.460 | 0.343 | 0.223 | 0.766 | 0.298 | 0.331 |

an assessment of the reliability. The following table gives the results of 5 runs of 1000 ways of random splitting the data from School A :

| *Run* | *T-value* | *p-value* |
|---|---|---|
| 1 | -0.5550 | 0.5790 |
| 2 | 0.4025 | 0.6874 |
| 3 | -0.9992 | 0.3179 |
| 4 | -0.2248 | 0.8222 |
| 5 | -1.0748 | 0.2827 |

High *p*-values indicate strong reliability.

## 5.4 Validity of RRT

### *Exploratory Factor Analysis*

Assessment of construct validity can be measured by Factor Analysis, where the researcher allows the data to determine the factor structure. We use the data of School D. The analysis produces three significant eigen, values, the resulting loading matrix produces following groupings we find that

- Q4 and Q8 are treated within F1
- Q1, Q5 and Q7 are treated within F2

- Q2, Q6 and Q3 are treated within F3.

So a clustering algorithm was used which provides the natural four groups of Q1 and Q5, Q2 and Q6, Q3 and Q7 and Q4 and Q8.

***Confirmatory Factor Analysis :***

A structural model of the form

$$\eta = B\eta + \Gamma\xi + \zeta,$$

where $\eta$ is the response vector and $\xi$ represents the latent variable and $\zeta$ represents the random disturbances. Following two models have been fitted :

- Model 1 : Latent Variable 1 : Q1, Q2, Q5, Q6
  Latent Variable 2 : Q3, Q4, Q7, Q8
- Model 2 : Latent Variable 1 : Q1, Q2, Q5, Q6
  Latent Variable 2 : Q3, Q4, Q7, Q8
  Latent Variable 3 : Q5, Q6
  Latent Variable 4 : Q3, Q4

These models produce chi-square values of 18.06 and 4.43 with respective *p*-values 0.32 and 0.83, which indicate no significant departure from the models and provide high validity. Other criteria also point to the high validity of RRT.

### *5.5 Reliability and Validity of GRT*

#### *5.5.1 Reliability of GRT*

Three measures have been used to test the reliability of GRT.

***Percentage of Rating***

Percentage of agreement is common measure of reliability of ratings. In School B, for administering the GRT, each question was repeated twice. Percentage of agreement from the two records is given below :

| Q1 | Q2 | Q3 | Q4 |
|---|---|---|---|
| 0.818 | 0.857 | 0.792 | 0.961 |

*Adjusted percentage of Agreement*

Correction of the above measure for random agreements may be given as follows :

$$\frac{P_0 - P_e}{1 - P_e},$$

where $P_e = \sum_i P_{i.}p_{.i}$ and $P_{ij}$ refer to percentage agreement for the $j^{th}$ group at the $i^{th}$ occasion, $P_0$ is the total observed percentage of agreement. This is summarized below :

| Q1 | Q2 | Q3 | Q4 |
|---|---|---|---|
| 0.47 | 0.56 | 0.50 | 0.92 |

*Intra-class Correlation Coefficient **(ICC)***

The following table shows the numbers of positive group responses in the two runs of GRT.

| Run | Q1 | Q2 | Q3 | Q4 |
|---|---|---|---|---|
| 1 | 61 | 65 | 53 | 33 |
| 2 | 59 | 58 | 55 | 32 |

The intra-class correlation is found to be 0.91.

*5.5.2 Validity of GRT*

*External Validity*

After asking a sensitive question twice, the results may be

consistent or they are opposite. In the first case, the answer at the second occasion is used as a confirmatory test. In the second case, the sensitive question is asked again and the result of the third question is considered to be a result of confirmatory response. The result is presented in the following dichotomous table :

| | | Confirmatory | | Total |
|---|---|---|---|---|
| | | + | – | |
| GRT | + | A | B | A + B |
| | – | C | D | C + D |
| | Total | A + C | B + D | N = A + B + C + D |

Specificity and sensitivity are used as measures of validity, which are defined as follows :

$$\text{Specificity} = \frac{D}{B+D}$$

$$\text{Sensitivity} = \frac{A}{A+C}$$

These are given in the following table :

### (a) *Sensitivity*

| Q | 1st Case | 2nd Case | Average |
|---|---|---|---|
| Q1 | 0.933 | 0.950 | 0.942 |
| Q2 | 1.000 | 0.949 | 0.975 |
| Q3 | 0.923 | 0.962 | 0.942 |
| Q4 | 0.969 | 1.000 | 0.984 |

### (b) *Specificity*

| Q | 1st Case | 2nd Case | Average |
|---|---|---|---|
| Q1 | 0.706 | 0.882 | 0.794 |
| Q2 | 0.800 | 0.800 | 0.800 |
| Q3 | 0.667 | 0.889 | 0.778 |
| Q4 | 0.956 | 1.000 | 0.978 |

### *Criterion Validity*

In the department of Geography of School B, there were 34 students which were interviewed both by GRT and RRT. The consistency as measured by the ICC may then be used as a measure of validity. The ICC computed on the basis of these estimates is obtained to be 0.45.

## 6. Discussion and Summary

Group Response Technique is a valid alternative to Randomized Responses Technique for estimating population prevalence with respect to sensitive questions. If batches are not small, the privacy of responses are totally protected. This technique can be used with mail-in questionnaires by collecting the responses in batches; however, this may not guarantee a good response rate as the respondents may not be assured of their anonymity.

Thus, GRT has a serious limitation that it should be performed in a group setting where the interviewer should demonstrate the complete working of the experiment such as in a class room setting. The other limitation is that the batches must be of equal size. The mathematical formulation breaks down if this is not ascertained.

About RRT, it is thought that a significant proportion of the respondents may be skeptical of the procedure; however, with GRT this aspect may be considered to be minimized. We have provided a working solution for Bayes procedure of estimation, which has desirable properties even using classical measures when the prior is not very informative.

It is found that alternative form reliability of RRT is moderate ranging from 0.63 to 0.98, but the ICC is pretty high. The simulation by split halves supports the stability of RRT estimates. Also the construct validity of RRT is found to be satisfactory using the exploratory and confirmatory data analysis.

GRT is found to have high reliability as measured by ICC as well as percentage agreement at two occasions. The adjusted percentage is not very high but sensitivity and specificity are very high. It is conjectured that the reliability of RRT may be more variable than that of GRT. The criterion validity measure in GRT is low (0.45) but we do not have such a measure available for RRT.


## Acknowledgements

Some parts of this paper are based on research work carried out by Weiming Li for his thesis under the direction of the author. Partial support by a research grant from NSERC of Canada is also acknowledged.


## References


Abramowitz, A.M. and Stegun, I.A. (1960), *Handbook of Mathematical Functions with Formulas, Graphs and Mathematical Tables*, Bureau of National Standards, Washington, D.C., USA.

Boswell, M.T. and Patil, G.P. (1987), A Perspective on Composite Sampling, *Communications in Statistics – Theory and Methods*, 16, 3069-3093.

Boswell, M.T., Gore, S.D., Patil, G.P. and Taillie, C.(1992), A Bayesian Approach to Classifying Samples as Polluted or Non-polluted, *Technical Report* No. 92-0801, State College, PA : Department of Statistics, The Pennsylvania State University.

Chaubey, Y.P. and Li, W. (1995a), Comparison Between Maximum Likelihood and Bayes Methods for Estimation of Binomial Probability with Sample Compositing, *Journal of Official Statistics.*

Chaubey, Y.P. and Li, W. (1995b), Group Response Technique: An Alternative to Randomized Response Technique, *Technical Report* No. 5/95, Montreal, QC : Department of Mathematics and Statistics, Concordia University.

Chaubey, Y.P., Dwivedi, T.D. and Li, W. (1997), Batch Sampling as a Tool in Estimating Quality, *J.Nat. Acad. Math.*, 11, 128-141.

Garner, F.C., Stapanian, M.A. and Williams, L. (1987), Composite Sampling for Environmental Monitoring. In *Principles of*

*Environmental Sampling*, Ed. L.H. Keith, Washington, D.C.: American Chemical Society.

Garner, F.C., Stapanian, M.A., Yfantis, E.A. and Williams, L. (1989), Probability Estimation with Sample Compositing Techniques, *Journal of Official Statistics*, 5, 365-374.

Greenberg, B.G., Abernathy, J.R. and Horvitz, D.G. (1986), Randomized Response, *Encyclopedia of Statistical Science*, Eds. Kotz, S. et al., Wiley and Sons, 540-548.

Greenberg, B.G., Abul-Ela, A.A., Simmons, W.R. and Horvitz, D.G. (1969), The Unrelated Question Randomized Response Model : Theoretical Framework, *Journal of the American Statistical Association*, 64, 520-532.

Hammick, P.A. and Gastwirth, J.L. (1994), Group Testing for Sensitive Characteristics : Extension to Higher Prevalence Levels, *International Statistical Review*, 62, 319-331.

Horvitz, D.G., Shah, B.V. and Simmons, W.R. (1967), The Unrelated Randomized Response Model, *Proc. Soc. Stat. Sect. Amer. Stat. Assoc.*, 65-72.

Li, W. (1995), A Reliability and Validity Study of Randomized Response Technique and its Alternative : Group Response Technique, *Ph.D. Thesis* of W. Li, Concordia University, Montreal, Canada.

Liu, P.T. and Chow, L.P. (1976), The Efficiency of the Multiple Trial Randomized Response Technique, *Biometrics*, 32, 607-618.

Pitz, G.F. (1980), Bayesian Analysis of Randomized Response Model, *Psychological Bulletin*, 87, 209-212.

Warner, S.L. (1965), Randomized Response: A Survey Technique for Eliminating Evasive Answer Bias, *Journal of American Statistical Association*, 60, 65-69.

Spurrier, J.D. and Padgett, W.J. (1980), The Application of Bayesian Techniques in Randomized Response. In *Sociological Methodology*. Ed. K.F. Schuessler, San Francisco : Jossey-Bass, 533-544.

Winkler, R.L. and Franklin, L.A. (1979), Warner's Randomized Response Model : A Bayesian Approach, *Journal of the American Statistical Association*, 74, 207-214.

2

# Post-partum Amenorrhoea and Lactation : A Statistical Study

*H. J. Vaman* and *B. P. Madhura*

## 1. Introduction

The health, demographic, and economic benefits of breastfeeding extend from the individual level of the mother and the child to the family, the community and to the national level. At the individual level, breastfeeding provides superior nutrition, helps in developing resistance to infections, and thus the baby becomes less likely to develop allergies. Also, breastfed babies have fewer stomach upsets and no constipation. Breastfeeding contributes to proper jaw and mouth development. Breastfed infants tend to have fewer incidences of cavities and need fewer orthodontic corrections later in life. Such babies also benefit emotionally, and are smarter because of better brain development. Most breastfed babies also cry less. Though these benefits, mentioned above, are from the infant's point of view, we do have concrete facts from several studies, which point to benefits of lactation to the mother. Some of these are listed below :

- Mothers who breastfeed, usually, lose weight more easily than mothers who feed formula to their babies.
- Also, the sucking action by the baby causes a mother's uterus to contract, which, in turn, reduces the flow of blood after delivery.

- Lactating mothers are less likely to develop breast cancer.
- The hormones a mother's body releases when she breastfeeds produces feelings of warmth, love and calmness. These help the mother to relax.

No multinational corporation, no government, no power structure of any kind can provide such a perfect baby food and all the other benefits. The consequent benefits accruing to the society as a whole are also enormous and go beyond the individual mother and her child. A rather surprising indirect benefit is derived by way of reduction in environmental pollution of air, water and land that could result from the production of artificial baby milk and its packaging. It does not use up water and energy resources for preparation of formula and for heating bottles. A study by the University of Michigan Health System reports a saving of about $2000 a year per family in the US.

Research reports have also revealed that mortality rate is low among infants who are breastfed than among those who are not. Also, infant feeding choices have a significant financial impact. Studies in the developed countries have estimated that average additional health care costs of a formula-fed infant, over those of a breastfed, were of the order of $1,400 for the first year alone. Thus, the annual savings in expenditure on formula and additional health care bills would be of the order of $2.4 billion if 1 million additional babies were breastfed every year (www.promom.org / bf_info).

Numerous fertility studies have revealed that post-partum ammenorrhoea period (duration between delivery and the subsequent onset of menstruation) is longer for women who breastfeed than those who do not. This role of breastfeeding as a contraceptive, especially during post-partum period, is very much evident in developing countries.

Certain studies that reveal the impact of the demographic

characteristics at the level of the community may be quoted as follows. Bongaarts (1978) observed that in Bangladesh, absence of lactation would have resulted in doubling of the population that one actually observed then. Breastfeeding provides as much as 35 million couple years of protection as compared to 27 million by the aggregate of all family planning programmes. The study of effect of duration and level of breastfeeding on the duration of post-partum amenorrhoea (PA) will help us in quantifying the contraceptive effects of different durations of levels of breastfeeding.

A stochastic model of the above problem at the level of the individual mother, with changing breastfeeding levels constituting an absorbing semi-Markov process, is proposed in this paper by considering the time spent by a woman in each of the breastfeeding states as sojourn times and considering that the sojourn time between two breastfeeding levels depends only on the adjoining breastfeeding states, leading to an embedding Markov chain. The development of the model and the statistical analysis based on it parallels Dinse and Larson (1986), and Lagakos et al. (1978). Data from a field study are analyzed in this setting.

The statistical modelling and analysis of the effect of breastfeeding on the duration of ammenorrhoea following childbirth, familiar to the present authors, revolve around regression analysis, life-table approach, proportional hazards, and some are based on distributional models. With perhaps the exception of Habitch (1985) and Ginsberg (1974), these studies do not consider the role of the *level* of breastfeeding. Habitch's study is from a physiological perspective, while the latter provides a novel stochastic model that incorporates full and partial breastfeeding.

The second author of the present paper, while working for the project 'Community based assessment of lactational ammenorrhoea' found, as is widely believed, that the practice of switching from full breastfeeding to partial breastfeeding to be very common. An exploratory regression analysis showed that age and parity had little effect while the educational level had an impact on the duration of PA. This and the review of the several studies prompted us to develop a model that accommodates switching the breastfeeding level, avoids the bias due to self-selected sub-samples and leads to valid estimates.

This paper gives a multistate model based on a semi-Markov process with the different breastfeeding levels as the states, switching from one level to another as a transition, and the duration of breastfeeding at a level as a sojourn time. No distributional or continuity assumption for the sojourn times is made. The model is employed on a real data set, estimates are obtained, and certain general conclusions are made.

## 2. The Model

In what follows, we model the breastfeeding pattern of a lactating mother from the time of parturition to the resumption of the menstruation cycle, on the lines of the model in Lagakos et al. (1978). Let X(t) denote the level of breastfeeding at time t for a woman. Change from one level of breastfeeding to another, or to the Non Amenorrhoeic State (NAS) is considered to be an *event*. It is assumed that state NAS would be reached when menstruation returns, although in reality this happens when ovulation resumes. The process {X(t)} is thus considered to be an absorbing stochastic process transiting through different breastfeeding (bf) levels and terminating at the NAS. The state space is S = {0,1,2,...,(s-1), s}, and in the present application, state '0' is the no breastfeeding state, state 1 is for complete breastfeeding (cbf), states 2 to (s - 1) are levels of partial breastfeeding (pbf), and s = NAS. States 1 to (s – 1) are transient and s is the absorbing state. Suppose that $F_{ij}(.)$ denotes the distribution function of duration of bf at level i before shifting to level j. Thus it gives the sojourn time distribution at level i for a switch to level j. In the present situation, the assumption of $F_{ij}(.)$ being exponential is untenable. Letting $F_{ij}(.)$ to be arbitrary renders the process semi-Markov. We however assume that the process is time homogeneous.

Let $T_n$ denote the sojourn time, duration of bf, between the (n-1)$^{st}$ and the n$^{th}$ epochs, X(t) jumps from state $X_{n-1}$ to state $X_n$. Given the initial level of breastfeeding, the process is characterized by the transition probabilities

$$p(i,j) = P(X_{n+1} = j \mid X_n = i),\ i, j \in S, \qquad \text{... (2.1)}$$

and the sojourn time distributions determined by

$$F_{ij}(t) = P(T_{n+1} \leq t \mid X_{n+1} = j, X_n = i),\ \ i, j \in S. \qquad \text{... (2.2)}$$

Evidently, it is assumed, to simplify the matters, that the distributions are invariant with respect to n.

### 3. Statistical Inference

Suppose that there are N mothers in the study and the nursing history of the i$^{th}$ woman recorded for m epochs is summarized by the data segment

$$D_i = \{X_0, T_1, X_1, T_2, \ldots, X_{m-1}, T_m, X_m\}, \qquad \ldots \ (3.1)$$

$1 \leq X_l \leq (s\text{-}1), 0 \leq l \leq (m\text{-}1)$ and $X_m = s$ if the i$^{th}$ woman has reached NAS. Since s is an absorbing state, a data segment $D_i$ is said to be complete if $\lambda_m = s$ for the i$^{th}$ woman, if not, the data segment is said to be censored, and we write $X_m = s + 1$. Thus,

$$X_m = \begin{cases} s, & \text{if } D_i \text{ is complete} \\ s+1, & \text{if } D_i \text{ is right censored,} \end{cases} \qquad \dots (3.2)$$

i = 1,2,3,..., N. The data output from the study will thus be D = {$D_i$, i = 1, 2, 3,...,N}, with the distribution of $T_n$ as in (2.2), the survival function of a sojourn time in state i, under the homogeneity assumption, is given by, S(t;i) = P($T_{n+1}$ > t | $X_n$ = i), i =1,2,...,s. In other words, S(t;i) gives the probability that the woman practicing i[th] breastfeeding level continues to do so beyond time t.

We now reparametrize the model, as in Prentice et al. (1978), in terms of the 'hazard rate' functions specific to the transition at a given time, defined in (3.3) and maximize the likelihood. This approach leads to nonparametric estimators of the Kaplan-Meier type.

Let $\tau_1, \tau_2, \dots, \tau_r$ denote r ($\leq$ N) distinct, and ordered (increasing) uncensored sojourn times (durations of breast feeding) observed across all women and all bf levels with $m_{ijk}$ as the multiplicity of $\tau_k$ among transitions from state i to state j, i = 1,2,3,...(s-1), j = 1, 2,..., s, and k = 1,2,3,...r. Thus, $m_{ijk}$ gives the number of women in the study group switching from bf level i to state level j after spending $\tau_k$ 'months', in particular, $m_{i(s+1)k}$ denotes the number of mothers who are still lactating at bf level i for a duration that lies in $[\tau_k, \tau_{k+1})$, and continue to be amenorrhoeic at the time of estimation, a censored observation, k = 0,1,2,..., r. It is convenient to take $\tau_0 = 0$ and $\tau_{m+1} = \infty$.

Assuming that the distribution of $T_n$ is discrete with positive mass only at $\tau_1, \tau_2, \dots, \tau_r$ and also over the open interval $(\tau_{r+1}, \infty)$ if the largest sojourn time is a censored one, the process X(t) described by p(i,j) and $F_{ij}(.)$, vide (2.1) and (2.2) above, can also be reparametrized in terms of the transition and sojourn time specific hazard functions $\{\lambda_{ijk}\}$, which are functions of p(i,j) and $F_{ij}(.)$, and are defined by

$$\lambda_{ijk} = P(T_{n+1} = \tau_k, X_{n+1} = j \mid T_{n+1} \geq \tau_k, X_n = i) \qquad ...(3.3)$$

i =1, 2,..., s-1; j = 1, 2,...,s, k = 1,2,...,r. Thus, $\Sigma\lambda_{ijk}$ is a hazard function corresponding to $\tau_k$.

The assumption that the distribution of $T_n$ is discrete makes S(t;i) a step function with jumps at $\tau_1, \tau_2,..., \tau_k$, $S(t;i) = S(\tau_k;i)$, $\tau_k \leq t < \tau_{k+1}$, $S(\tau_k;i) = P(T_{n+1} > \tau_k \mid X_n = i)$. Let $q_{ik}$ denote the conditional probability that a mother, breast-feeding at level i, continues to lactate beyond $\tau_k$, without getting back her menses given that she has already done so for $\tau_{k-1}$ months. These definitions yield

$$q_{ik} = \frac{S(u_k;i)}{S(u_{k-1};i)} \qquad ... (3.4)$$

$$= 1 - \sum_{j=1}^{s} \lambda_{ijk},$$

and for $0 \leq t < \tau_k$, and k = 0, 1,..., r,

$$S(t;i) = \frac{S(u_1;i)\, S(u_2;i) \ldots S(u_k;i)}{S(u_0;i)\, S(u_1;i) \ldots S(u_{k-1};i)},$$

$$= \prod_{l=1}^{k} q_{il}. \qquad ... (3.5)$$

Let $D^c$ and $D^{rc}$ represent the collection of complete and right censored data segments, respectively. It may be noted that the data segments are independent histories pertaining to N different mothers.

Note that $\sum_{l=k}^{r} \sum_{j=1}^{s} m_{ijl}$ is the total number of mothers still nursing at level i but who would enter the NAS state before the end of study obtained from $D^c$ and the sum $\sum_{l=k}^{r} m_{i(s+1)l}$ is the total number of mothers who are still nursing at level i and who would remain amenorrhoeic at the end of the study obtained from $D^{rc}$. Then,

$$n_{ik} = \sum_{r=k}^{m} \sum_{j=1}^{s} m_{ijr} + \sum_{r=k}^{m} m_{i,(s+1),r}$$

gives the total number of women still breastfeeding at level i beyond $\tau_{k-1}$ months.

The initial probabilities do not play any role in conditional inference. The likelihood of $\lambda = (\lambda_{ijk})$ under the discretised setup can now be expressed as

$$L(\lambda | D) = L(\lambda | D^c)\, L(\lambda | D^{rc}),$$

with

$$L(1/D) a \prod_{k=1}^{r} \prod_{i=1}^{s} \prod_{j=1}^{s} (\lambda_{ijk})^{m_{ijk}}$$

and

$$L(\lambda \mid D^{rc}) \propto \prod_{k=1}^{r} \prod_{i=1}^{s} (1 - \sum_{j=1}^{s} \lambda_{ijk})^{n_{ik} - \sum_{j=1}^{s} m_{ijk}}.$$

Then the log likelihood function of D, $l(\lambda \mid D)$, is

$$l(\lambda | D) = \sum_{k=1}^{r} \sum_{j=1}^{s} [ (n_{ik} - \sum_{j=1}^{s} m_{ijk}) \log(1 - \sum_{j=1}^{s} \lambda_{ijk}) + \sum_{j=1}^{s} m_{ijk} \log \lambda_{ijk} ].$$

This yields MLE of $\lambda_{ijk}$ as $m_{ijk} / n_{ik}$. Employing this estimator of $\lambda_{ijk}$ in appropriate expressions, estimators of $q_{ik}$, S(t;i), $p_{ijk}$, p(i,j) and $F_{ij}(t)$ are obtained. More on this can be found in the paper of Lagakos et al. (1978).

We may also test certain hypotheses of interest, like, the distribution of the duration of a partial breastfeeding level is the same as the distribution of the duration of complete breastfeeding, i.e.,

$$H_1 : F_{i1}(t) = F_{i2}(t) = \cdots = F_{i(s-1)}(t),$$

or, that the distribution of reaching NAS (the absorbing state) from different breastfeeding levels are homogeneous, i.e.,

$$H_2 : F_{1s}(t) = F_{2s}(t) = \ldots = F_{(s-1)s}(t).$$

For testing these hypotheses, we may use the large sample test based on the –2log (*LR*) statistic which has the same form as the one for testing homogeneity in a contingency table as shown by Dinse and Larson (1986).

## 4. Application to Real Data

In a follow up study conducted from September 1992 to August 1993, titled 'Community based Assessment of Lactational Amenorrhoea' sponsored by the Indian Association of Preventive and Social Medicine data were collected on 137 mothers in a Public Health Centre near Bangalore. Information on mothers' feeding habits, their amenorrhoeic status along and on certain socio-demographic variables were collected on a monthly basis. Results of application of the proposed model to a real life data set are as follows.

Let us define the following parameters with reference to the proposed model :

- Est $\lambda_{CPt}$ = estimated 'hazard' rate for mothers in cbf who shift to pbf after t months
- Est $\lambda_{CAt}$ = estimated 'hazard' rate for mothers in cbf who shift to absorbing state after t months
- Est $\lambda_{PAt}$ = estimated 'hazard' rate for mothers in pbf who get into the NAS after t months.

Estimates for the above quantities for the data are given below :

| t | Est $\lambda_{CPt}$ | Est $\lambda_{CAt}$ | Est $\lambda_{PAt}$ |
|---|---|---|---|
| 1 | 0 | 0 | 0 |
| 2 | 0.016 | 0.024 | 0.330 |
| 3 | 0.035 | 0.027 | 0.250 |
| 4 | 0.026 | 0.038 | 0.429 |
| 5 | 0.261 | 0.029 | 0.583 |
| 6 | 0.426 | 0.043 | 0.488 |
| 7 | 0.739 | 0.174 | 0.648 |
| 8 | 0.438 | 0.125 | 0.778 |
| 9 | 0.438 | 0.188 | 0.830 |
| 10 | 0.800 | 0.200 | 0.872 |
| 11 | 0 | 0 | 0.833 |
| 12 | 0 | 0 | 0.913 |

These estimates can be used in equations (3.4) and (3.5) to obtain estimates of the curve S(i, t) for different states. Following hypotheses were tested at 3, 6, 9 and 12 months :

- $H_0$ : Time spent in any transient breastfeeding state is independent of the next event
  $H_1$ : F(3;C,P) = F(3;C,A); $H_1$ : F(6;C,P) = F(6;C,A); $H_1$ : F(9;C,P) = F(9;C,A);

The tests resulted in rejection of the hypotheses at 3 and 6 months, a borderline case at 9 months :

- $H_0$ : Sojourn time distribution of different transient

breastfeeding states before reaching absorbing state are homogeneous
$H_2 : F(3;C,A) = F(3;P,A); H_2 : F(6;C,A) = F(6;P,A); H_2 : F(9;C,A) = F(9;P,A)$.

Here, the result was to reject the hypotheses at 3, 6 and 9 months revealing that the probability distributions of reaching absorbing state from CBF and PBF are not homogeneous.

## Acknowledgement

The authors wish to acknowledge Dr. Ashwini Mathur for providing with some of the useful references in this area.

## References

Bongaarts, J. (1978), A Framework for Analyzing the Approximate Determinants of Fertility. *Pop. and Dev. Rev.*, 105-132.

Dinse, G. N. and Larson, M.T. (1986), A Note on Semi-Markov Models for Partially Censored Data,. *Biometrika*, 73,379-386.

Ginsberg, R.B. (1974), The Effect of Lactation on Post-partum Anovulatory Period : an Application of a Bivariate Model. *Theoretical Biology*, 4,276 -299.

Habitch, J.P. (1985), The Contraceptive Effect Role of Breastfeeding, *Population Studies*, 39,213-232.

Lagakos S.W., Sommer, C.J., and Zelen, M. (1978), Semi-Markov Models for Partially Censored Data, *Biometrika*, 65,111-117.

Prentice, R.L., Kalbfleisch, J.D., Peterson, A.V., Flournoy, N., Farewell, V.T., and Breslow, N.E. (1978), The Analysis of Failure Times in the Presence of Competing Risks, *Biometrics*, 34, 541-554.

UMHS : News Bulletin in University of Michigan Health System web site. *(http://www.med.umich.edu/1libr/womens/breast05.htm), www.promom.org/bf_info.*

# 3

# A New Measure of Occupational Mobility

*Asis Kumar Chattopadhyay*

## 1. Introduction

In recent years there has been growing interest in the studies of manpower planning where occupational mobility plays a central role in manpower planning. In broader term, it refers to the movement of employees between jobs. As a consequence of globalization and expansion of job market in the non-traditional sectors, the phenomenon of changing jobs has gained greater attention of the researchers and planners. Unlike social mobility (Prais,1955), there is no fixed time interval between successive moves in occupational mobility. Hodge (1966) studied the occupational mobility as a probability process. Stewman (1975) discussed the occupational mobility using Markov model. A comprehensive summary of the theoretical developments and practical applications of the occupational mobility have been provided by Stewman (1976). Ginsberg (1971, 1972a, 1972b, 1978, 1979a, 1979b) has made several attempts to describe the occupational mobility patterns in terms of semi-Markov processes.

Bartholomew (1982) suggested measures of occupational mobility based on the matrix of transition probabilities and the stochastic process $m(t)$, where $m(t)$ is the random number of time points at which individuals decide to change their existing job during the interval $(0,t)$. Mukherjee and

Chattopadhyay (1985) developed a measure by considering successive changes in occupation of an individual as constituting a renewal process. Later Mukherjee and Chattopadhyay (1989) proposed a measure based on the reward structure. Chattopadhyay and Baidya (1999) considered salary based occupational categories to study social mobility. Khan and Chattopadhyay (2003) developed a predictive measure of occupational mobility based on number of job offers. None of the measures available in the literature has taken into account the explicit role of the reward (remuneration) associated with job offer that directly influences the pattern of job changes.

In this study we propose a new measure of occupational mobility based on the number of job changes and the associated wages. The distribution of the total number of job changes during an individual's entire service life up to the point of study has been derived by considering times between consecutive job offers (within an interval) and the corresponding wages to be independent random variables. The proposed measure of occupational mobility based on the number of job offers and wage distribution has been suggested primarily in the general setup and then its value has been derived for special situations where some specific assumptions regarding the number of job offers and wage distributions have been made. A survey has been conducted among the staff members of the University of Southern Queensland, Australia to collect data on occupational mobility. The proposed model has been fitted to the survey data to explain the occupational mobility pattern among the employees of the University. For this particular application of the occupational mobility model the number of job offers has been found to best fit the geometric distribution, and the wages best fit a gamma distribution.

## 2. The Model

Let the service life of an individual be comprised of $k$ intervals of fixed width, $t$. It is assumed that the individual gets at least one job offer within each such interval. The worth of an offer being determined by the associated salary (reward). The individual (assumed to be in service already) decides to leave the present job or not, at the end of each interval. One moves to a new job for the first time at the end of an interval in which the maximum of the remunerations associated with different job offers (within that interval) exceeds a fixed amount. This is the minimum wage at which the individual is willing to enter the job market for the first time. Subsequently, one changes the current job at the end of a particular interval only when the maximum of the wages associated with the offers received during that interval exceeds the wage of the current job. A change of job in this paper means that an individual may move from one occupation to another or within the same occupation. Let the individual gets $N_i$ new job offers in the $i^{th}$ interval and let $X_{ij}$ be the salary corresponding to the $j^{th}$ job offer in the $i^{th}$ interval, for $j = 1, 2, \ldots, n_i$, and $i = 1, 2, \ldots, k$. Note that to reflect the real life situation it is necessary to assume that $n_i$ is strictly greater than zero, since none can enter into the job market without a job offer. Both $X_{ij}$ and $N_i$ are assumed to be independently and identically distributed with pdf $g(x)$, $0 < x < \infty$, and pmf $h(y)$, $y = 1, 2, \ldots, \infty$, respectively. Define

$$Z_i = \max(X_{i1}, X_{i2}, \ldots, X_{in_i}). \qquad \ldots (2.1)$$

Here $Z_i$ is the maximum wage of all job offers during the $i^{th}$ interval. Since $Z_i$ is the largest order statistic, for a given $n_i$, the pdf of the conditional distribution of $Z_i$ is

$$f(z_i \mid n_i) = n_i[G(x_{ij})]^{n_i - 1} g(z_i),$$

where $G(\cdot)$ is the cdf of the distribution of $X_{ij}$. The pdf of the joint distribution of $Z_i$ and $N_i$ becomes

$$f(z_i, n_i) = n_i[G(x_{ij})]^{n_i-1}g(z_i)h(n_i).$$

Hence the marginal distribution of $Z_i$ is

$$f(z_i) = \sum_{n_i=1}^{\infty} n_i\,[G(z_i)]^{n_i-1}g(z_i)h(n_i), \qquad \text{... (2.2)}$$

where g (.) and h (.) have the same specifications as before. Let $F_{Z_i}(z)$ denote the corresponding cdf. Let $z_0$ be the minimum wage for which the individual accepts the first job offer at the $i^{th}$ interval. Then we can define

$$F_{Z_i}(z_0) = P[Z_i < z_0] \qquad \text{... (2.3)}$$

and its complement

$$\overline{F}_{Z_i}(z_0) = 1 - F_{Z_i}(z_0) = P[Z_i > z_0]. \text{ ... (2.4)}$$

## 3. Distribution of the Number of Job Changes

In this section we derive the distribution of the number of job changes during the service life of an individual. Define $N(k)$ = total number of job changes within the service life of the individual and $P_r^{(k)}$ = the probability of $r$ job changes in the entire service life of the individual. Then

$$P_r^{(k)} = P[N(k) = r]. \qquad \text{... (3.1)}$$

**Theorem 3.1 :** Under the above definition of $F_{Z_i}(z_0)$ and $P_r^{(k)}$, we have

$$P_r^{(k)} = F^k \qquad \text{if } r = 0$$

$$= F^{k-1}\overline{F}\,[\textstyle\sum_{m=0}^{k-r}\binom{r+m-1}{m}(2F)^{-(r+m-1)}] \text{ if } 1 \leq r \leq k, \text{... (3.2)}$$

where, for notational convenience, we write $F = F_{Z_i}(z_0)$ and $\overline{F} = 1 - F_{Z_i}(z_0)$.

**Outline of proof :**

Note that $P[Z_i > Z_j] = P[Z_i < Z_j] = 0.5$ for $i, j = 1, 2, \ldots, k$, but $i \neq j$; and $P[Z_i > \max(Z_1, Z_2, \ldots, Z_{i-1})] = P[Z_i < \max(Z_1, Z_2, \ldots, Z_{i-1}] = 0.5$. ... (3.3)

Now define $S_i$ as the event that there is a job change in the $i^{th}$ time interval which depends only on the maximum wages of the $i^{th}$ and $(i-1)^{th}$ time intervals, for $i = 1, 2, \cdots, k$ and $T_i$ as the event that there is a job change in the $i^{th}$ interval which depends on the maximum wages of all intervals up to the $i^{th}$ including $z_0$, the initial minimum acceptable wage, for $i = 2, 3, \cdots, k$, that is

$S_i = [z_i > z_{i-1}]$, for $i = 1, 2, \ldots, k$ ... (3.4)
and $T_i = [z_i > max\,(z_0, z_1, \ldots, z_{i-1})]$, for $i = 2, 3, \ldots, k$. ... (3.5)

Then $P_r^{(k)}$ can be obtained by adding the probabilities of the $(k - r + 1)$ events $E_0, E_1, \ldots, E_{k-r}$, where

- $E_0 =$ there is no change in the first $(k - r)$ intervals and r changes in the last $r$ intervals,
- $E_1 =$ there is no change in the first $(k\text{-}r\text{-}1)$ intervals, one change at the $(k\text{-}r)^{th}$ interval and $(r-1)$ changes among the last $r$ intervals,
- $E_2 =$ there is no change in the first $(k - r - 2)$ intervals, one change at the $(k-r-1)^{th}$ interval and $(r - 1)$ changes among the last $(r + 1)$ intervals, $\cdots$
- $E_{k-r} =$ there is a change at the first interval and $(r - 1)$ changes among the last $(k-1)$ intervals.

Let $S^c$ denote the complement of the event $S$. Then from the fundamental rule of probability we have

$$P(E_0) = P[\, S_1^c S_2^c \ldots S_{k-r}^c T_{k-r+1} S_{k-r+1} \ldots S_k \,]$$

$$= F^{k-r}\overline{F}(0.5)^{r-1} \qquad \ldots (3.6)$$

Similarly,

$$P(E_1) = F^{k-r-1}\overline{F}\binom{r}{1}(0.5)^r$$

$$P(E_2) = F^{k-r-2}\overline{F}\binom{r+1}{2}(0.5)^{r+1}$$

$$P(E_{k-r}) = \overline{F}\binom{r+(k-r-1)}{k-r}(0.5)^{r+(k-r-1)}. \quad \ldots (3.7)$$

Hence the proof is complete by adding the above probabilities.

*Illustration* : Consider the situation when $k = 3$ and $r = 2$.

$$P_2^{(3)} = P(S_1S_2S_3^c) + P(S_1S_2^c\ T_3\ ) + P(S_1^c\ T_2S_3\ )$$

$$= P(Z_1 > z_0)P(Z_2 > Z_1)P(Z_3 < Z_2) + P(Z_1 > z_0)P(Z_2 < Z_1)P(Z_3 > Z_1) + P(Z_1 < z_0)P(Z_2 > z_0)P(Z_3 > Z_2)$$

$$= \overline{F}(0.5)(0.5) + \overline{F}(0.5)(0.5) + F\overline{F}(0.5)$$

$$= (0.5)\overline{F}(1+F). \quad \ldots (3.8)$$

When $r = 0$,

$$P_0^{(3)} = P(S_1^cS_2^cS_3^c)$$

$$= P(Z_1 < z_0)P(Z_2 < z_0)P(Z_3 < z_0) = F^3. \quad \ldots (3.9)$$

**Theorem 3.2 :** Under the above setup $P_r^{(k)}$ is a probability distribution, i.e.,

$$\sum_{r=0}^{k} P_r^{(k)} = 1. \quad \ldots (3.10)$$

**Proof :** Write $s = r + m - 1$. Then

$$\sum_{r=0}^{k} P_r^{(k)} = F^k + F^{k-1}\overline{F}\left[\sum_{r=1}^{k}\sum_{m=0}^{k-r}\binom{r+m-1}{m}(0.5F)^{r+m-1}\right]$$

$$= F^k + F^{k-1}\overline{F}\left[\sum_{s=0}^{k-1}(0.5)^2\sum_{m=0}^{s}\binom{s}{m}\right]$$

$$= F^k + F^{k-1}\overline{F}\left[\sum_{s=0}^{k-1}(0.5F)^s 2^s\right] = 1. \qquad \text{... (3.11)}$$

## 4. Measure of Occupational Mobility

In this section we obtain a measure of occupational mobility using $P_r^{(k)}$ as defined in Section 3. From the previous specifications, the moments of the number of job changes are reasonable choices as measures of occupational mobility. For practical reasons, the first raw moment has better intuitive appeal in interpreting the phenomenon of job changes than any other moments. Therefore, we suggest, the expectation of the number of job changes can be considered as a measure of occupational mobility. This should of course be normalized with respect to $k$. Then we have

$$E[N(k)] = \sum_{r=0}^{k} rP_r^{(k)}$$

$$= F^{k-1}\overline{F}\sum_{r=1}^{k}\sum_{m=0}^{k-r} r\binom{r+m-1}{m}(0.5F)^{r+m-1}$$

$$= [(k+1)\overline{F} - \overline{F}F^k - F(1-F^k)]/2\overline{F} \qquad \text{... (4.1)}$$

In the computation of $E[N(k)]$ different binomial and geometric series are involved. After normalization with respect to $k$, the measure becomes $E[N(k)/k]$.

In a similar way it can be shown that

$$E[N(k)]^2 = (0.25)F^{k-1}\overline{F}\sum_{s=0}^{k-1}(1/F)^{s}(s^2+5s+4) \qquad \ldots (4.2)$$

Hence the variance of $N(k)$, $Var[N(k)]$ is readily available from (4.1) and (4.2). As a measure of spread of the above measure of occupational mobility one uses the estimated value of $Var[N(k)]/k^2$. Computing procedures of $E[N(k)]$ and $Var[N(k)]$ are given in Section 5.

## 5. Application

With a view to applying the proposed measure of occupational mobility a survey among the employees of the University of Southern Queensland (USQ), Australia has been conducted. The main objective of the survey is to gather data on the number of job offers received by the individual employees during the entire employment period, including the offer(s) of the current employer. In addition, wages associated with each of the job offers for the employees have been collected. The data on the number of job offers and the associated wages have been classified according to the staff category, academic and non-academic, as well as gender, male and female. Separate analysis of the data have been provided based on the above four categories and the values of the proposed measure of occupational mobility have been obtained for all those cases. An overall analysis of the data from all the respondents across the categories has also been provided.

The sample consisted of 221 employees of the USQ. This comprises of 83 academic and 138 non-academic staff. Among these respondents there are 93 males and 128 females. In the survey we limited the maximum number of offers for any individual to 20 and the range of wages has been equally divided into 6 intervals. In computation of the measure of occupational mobility and all associate functions as well as occupational mobility and all associate functions as well as fitting of different distributions we have used MATLAB and SPSS packages. Some Pascal programming has also been used for the fitting of the distribution.

### *5.1 The Survey*

The general distribution of the survey data with respect to the number of job offers and associated wages is provided in this sub-section. Table 3.1represents the means and standard deviations of the number of job offers of the respondents by various categories.

**Table 3.1** : Summary Statistics of Number of Job Offers by Gender and Staff Category

| | *Academic* | | | *Non-Academic* | | | *Total* | | |
|---|---|---|---|---|---|---|---|---|---|
| | *Count* | *Mean* | *Std.* | *Count* | *Mean* | *Std.* | *Count* | *Mean* | *Std.* |
| Male | 58 | 5.40 | 4.26 | 35 | 6.20 | 4.31 | 93 | 5.70 | 4.27 |
| Female | 25 | 6.20 | 4.07 | 103 | 5.60 | 4.24 | 128 | 5.72 | 4.20 |
| Total | 83 | 5.64 | 4.20 | 138 | 5.75 | 4.25 | 221 | 5.71 | 4.22 |

The average number of job offers received by the employees in the sample is 5.71. The corresponding figure for the academics is 5.64 and that of the non-academics is 5.75. Thus the average number of job offers for non-academics is slightly higher than the academics. The female respondents have higher average number of job offers than their male counterparts, although the difference is negligible. However, a clear dominance of the females with respect to the average number of job offers over the males is observed among the academics. The opposite picture is reflected in the case of non-academic staff. The value of the measure of spread remains almost the same across the categories. For the remaining parts of the paper the value of $t$ and $k$ are assumed to be 1 and 20, respectively, for computations.

### *5.2 Fitting of Distributions*

First we have fitted the geometric distribution with pmf

$$h(y) = p(1-p)^{y-1}, y = 1, 2, 3, \cdots, \infty \qquad \dots (5.1)$$

to the number of job offers for all respondents as well as by all the categories. Here the estimate of the parameter $p$ is $\hat{p} = 1/\bar{y}$, where $\bar{y}$ is the sample mean of the observed number of job offers.

From Table 3.2 it is evident that the data fit very well with the geometric distribution for all respondents as well as over all the categories.

**Table 3.2 :** Expected and Observed Frequency Distributions of Job Offers by Gender and Staff Category (for geometric distribution)

| *Offers* | *Academics* | | *Non-Acads.* | | *Male* | | *Female* | | *Total* | |
|---|---|---|---|---|---|---|---|---|---|---|
| | *Exp.* | *Obs.* | *Exp.* | *Obs.* | *Exp.* | *Obs.* | *Exp.* | *Obs.* | *Exp.* | *Obs.* |
| 1 | 15 | 12 | 25 | 24 | 16 | 15 | 22 | 21 | 36 | 39 |
| 2 | 12 | 9 | 20 | 12 | 13 | 8 | 18 | 13 | 21 | 32 |
| 3 | 10 | 11 | 17 | 16 | 11 | 14 | 15 | 13 | 27 | 26 |
| 4 | 8 | 6 | 14 | 12 | 9 | 7 | 13 | 11 | 18 | 22 |
| 5 | 7 | 11 | 11 | 12 | 8 | 9 | 10 | 14 | 23 | 18 |
| 6 | 6 | 5 | 9 | 12 | 6 | 8 | 9 | 9 | 17 | 15 |
| 7 | 5 | 7 | 8 | 9 | 5 | 6 | 7 | 10 | 16 | 12 |
| 8 | 4 | 6 | 6 | 9 | 4 | 6 | 6 | 9 | 15 | 10 |
| 9 | 3 | 3 | 5 | 9 | 3 | 5 | 5 | 7 | 12 | 8 |
| 10 | 3 | 4 | 4 | 6 | 3 | 3 | 4 | 7 | 10 | 7 |
| 11 | 2 | 2 | 3 | 1 | 2 | 1 | 3 | 2 | 3 | 6 |
| 12 | 2 | 1 | 3 | 4 | 2 | 3 | 3 | 2 | 5 | 5 |
| 13 | 1 | 0 | 2 | 4 | 2 | 2 | 2 | 2 | 4 | 4 |
| 14 | 1 | 2 | 2 | 3 | 1 | 2 | 2 | 3 | 5 | 3 |
| 15 | 1 | 0 | 2 | 1 | 1 | 0 | 2 | 1 | 1 | 3 |
| 16 | 1 | 2 | 1 | 0 | 1 | 2 | 1 | 0 | 2 | 2 |
| 17 | 1 | 0 | 1 | 1 | 1 | 0 | 1 | 1 | 1 | 2 |
| 18 | 1 | 1 | 1 | 1 | 1 | 1 | 1 | 1 | 2 | 1 |
| 19 | 0 | 0 | 1 | 1 | 1 | 0 | 1 | 1 | 1 | 1 |
| 20 | 0 | 1 | 1 | 1 | 0 | 1 | 1 | 1 | 2 | 1 |

Then we have fitted the gamma distribution with pdf

$$g(x) = \frac{1}{\beta^{\alpha}\,\Gamma(\alpha)} e^{-\frac{x}{\beta}} x^{\alpha-1}, 0 < x < \infty, \alpha > 0, \beta > 0 \quad \dots (5.2)$$

to the wages for all respondents as well as by all the categories. Salem and Mount (1974), and McDonald and Jensen (1979) used gamma distribution to model the distribution of income. The estimates of the parameters $\alpha$ and $\beta$ are

$\hat{\alpha} = \left[\frac{\bar{x}}{s_x}\right]^2$ and $\hat{\beta} = \left[\frac{s_x^{\,2}}{\bar{x}}\right]$, respectively, where $\bar{x}$ is the sample mean of the observed wages and $s_x$ is the corresponding standard deviation. For further details see Cohen and Whitten (1988). Angle (1992) used the two parameter gamma distribution to model the income distributions of blacks and the whites.

From Table 3.3 it is observed that the data fit more or less well to the gamma distribution for all respondents as well as over all the categories.

**Table 3.3 :** Expected and Observed Frequency Distributions of Wages by Gender and Staff Category (for gamma distribution)

| *Wages* | *Academics* | | *Non-Acads.* | | *Male* | | *Female* | | *Total* | |
|---|---|---|---|---|---|---|---|---|---|---|
| | *Exp.* | *Obs.* | *Exp.* | *Obs.* | *Exp.* | *Obs.* | *Exp.* | *Obs.* | *Exp.* | *Obs.* |
| 0 | 0 | 0 | 0 | 0 | 0 | 0 | 0 | 0 | 0 | 0 |
| 1 | 2 | 27 | 11 | 48 | 1 | 25 | 10 | 50 | 10 | 77 |
| 2 | 38 | 41 | 123 | 127 | 21 | 25 | 114 | 146 | 130 | 173 |
| 3 | 89 | 67 | 198 | 223 | 85 | 94 | 180 | 193 | 262 | 290 |
| 4 | 94 | 62 | 139 | 89 | 119 | 88 | 124 | 63 | 244 | 151 |
| 5 | 64 | 80 | 63 | 52 | 95 | 98 | 55 | 34 | 153 | 134 |
| 6 | 34 | 70 | 22 | 25 | 53 | 79 | 19 | 17 | 75 | 98 |

### 5.3 Computation of Measure of Occupational Mobility

Here we derive the expression for the measure of the occupational mobility under the above specifications of distribution of the number of job offers and wages. We also compute the values of the measure as well as the variance of the number of job changes during the employment period for all respondents. The same is also obtained for different categories of respondents to compare the mobility of employees over categories.

From (5.2), assuming $\alpha$ to be an integer (Evan et al. 1993, for instance), the pdf of the distribution of the maximum wage is

$$f(z_i) = \frac{1}{\beta^{\alpha}\Gamma(\alpha)} pe^{-\frac{z_i}{\beta}} z_i^{\alpha-1}\left((1-p)\left(1-e^{-\frac{z_i}{\beta}}\sum_{r=0}^{\alpha-1}\frac{1}{r!}\left\{\frac{z_i}{\beta}\right\}^r\right)\right)^{n_i-1} \quad \text{...(5.3)}$$

and the corresponding cdf is given by

$$F(z_0) = p\sum_{k=1}^{\infty} k(1-p)^{k-1}\tau_k(\alpha,\beta,z_0), \quad \text{... (5.4)}$$

where

$$\tau_k(\alpha,\beta,z_0) = \int_0^{z_0}\frac{1}{\beta^{\alpha}\Gamma(\alpha)} e^{-\frac{z_i}{\beta}} z_i^{\alpha-1}\left(1-e^{-\frac{z_i}{\beta}}\sum_{r=0}^{\alpha-1}\frac{1}{r!}\left\{\frac{z_i}{\beta}\right\}^r\right)^{k-1} dz_i. \quad \text{... (5.5)}$$

**Table 3.4 :** Empirical Distribution Function, Expectation, Variance of Job Changes and Measure of Mobility by Different Categories

| *Variable* | $\hat{\alpha}$ | $\hat{\beta}$ | $\hat{p}$ | $F(z_0)$ | $E[N(k)]$ | *Var [N(k)]* | $E\left[\frac{N(k)}{k}\right]$ |
|---|---|---|---|---|---|---|---|
| Academic | 6.54 | 0.61 | 0.1773 | 0.00060 | 10.49969 | 4.73451 | 0.52498 |
| Non Acad. | 6.38 | 0.48 | 0.1795 | 0.00200 | 10.49883 | 4.75000 | 0.52494 |
| Male | 8.23 | 0.50 | 0.1754 | 0.00020 | 10.49992 | 4.75000 | 0.52500 |
| Female | 6.00 | 0.48 | 0.1748 | 0.00400 | 10.49809 | 4.75000 | 0.52490 |
| Overall | 5.70 | 0.60 | 0.1751 | 0.00200 | 10.49896 | 4.75000 | 0.52495 |

Table 3.4 describes the expectation and variance of number of job changes for all respondents as well as categories and the values of the proposed measure. From the expected values it appears that for any particular

individual, on the average, the number of job offers with wage associated with a new job offer exceeding the maximum wage earned from a previous offer is a little over 10. The computed values of measure for the survey data enable us to infer that an employee is more mobile (than expected in the job market) in the occupational life depending on whether the observed value of the measure (computed on the basis of the employee's job changes) exceeds the above computed value of the measure. This is valid under the assumption that the job changes occur only on the basis of wage consideration and for specific values of $t$ and $k$. The same conclusion can be extended over all the categories considered in the study. The measure reveals the fact that the expected number of job changes is about the same for all employees regardless of the gender and the staff category.

**Acknowledgements**

The author is grateful to Prof. S.P. Mukherjee, Department of Statistics, Calcutta University for suggesting the problem. The author also offers sincere gratitude to the Department of Mathematics and Computing, Faculty of Sciences, University of Southern Queensland, Australia. Special thanks to all the participating staff of the USQ for responding to the survey.

**References**

Angle, J. (1992), The Inequality Process and the Distribution of Income to Blacks and Whites, *Journal of Mathematical Sociology*, 17(1), 77-98.

Bartholomew, D.J. (1982), *Stochastic Models for Social Processes*, Wiley, New York.

Chattopadhyay, A.K. and Baidya, K. (1999), Social Mobility among Residents of Calcutta, *Demography India*, 28, 215-224.

Cohen, A.C. (1991), *Truncated and Censored Samples*, Mercell Dekker, New York.

Cohen, A.C. and Whitten, B.J. (1988), *Parameter Estimation in Reliability and Life Span Models*, Mercell Dekker, New York.

Evans, M., Hastings, N. and Peacock, B. (1993), *Statistical Distributions*, Wiley, New York.

Ginsberg, R.B. (1971), Semi-Markov processes and mobility, *Journal of Mathematical Sociology*, 1, 233-262.

Ginsberg, R.B. (1972a), Critique of Probabilistic Models: Application of Semi-Markov Model to Migration, *Journal of Mathematical Sociology*, 2, 63-82.

Ginsberg, R.B. (1972b), Incorporating Causal Structure and Exogenous Information with Probabilistic Models : with Special Reference to Gravity, Migration and Markov Chains, *Journal of Mathematical Sociology*, 2, 83-103.

Ginsberg, R.B. (1978), The Relationship between Timing of Moves and Choice of Destination in Stochastic Models of Migration, *Environment and Planning*, 10, 667- 679.

Ginsberg, R.B. (1979a), *Stochastic Models of Migration* : Sweden 1961-1975, North Holland, New York.

Ginsberg, R.B. (1979b), Tests of Stochastic Models of Timing in Mobility Histories: Comparison of Information Derived from Different Observation Plans, *Environment and Planning*, 11, 1387-1404.

Hodge, R.W. (1966), Occupational Mobility as a Probability Process, *Demography*, 3, 19-34.

Khan, S. and Chattopadhyay, A.K. (2003), Predictive Analysis of Occupational Mobility, *Journal of Applied Statistical Science*, 12(1), 11-22.

McDonald, J. and Jensen, B.(1979), An Analysis of Some Properties of Alternative Measures of Income Inequality Based on the Gamma Distribution, *Journal of the American Statistical Association*, 47, 856-860.

Mukherjee, S.P. and Chattopadhyay, A.K. (1985), Renewal Theoretic Measurement of Occupational Mobility, *Calcutta Statistical Association Bulletin*, 34, 207-213.

Mukherjee, S.P. and Chattopadhyay, A.K. (1989), Measurement of Occupational Mobility using Semi-Markov Models, *Communications in Statistics-Theory and Methods*, 18(5), 1961-1978.

Prais, S.J. (1955), Measuring Social Mobility, *J. Roy. Statist. Soc.*, 118A, 56-66.

Salem, A. B. Z. and Mount, T. D. (1974), A Convenient Descriptive Model of Income Distribution : The Gamma Density, *Econometrica*, 42, 1115-1127.

Stewman, S. (1975), Two Markov Models of System Occupational Mobility : Underlying Conceptualization and Empirical Tests, *American Sociological Review*, 40, 298-321.

Stewman, S. (1976), Markov Models of Occupational Mobility : Theoretical Development and Empirical Support (Part 1), *Journal of Mathematical Sociology*, 4(2), 203-245.

4

# Exploratory Data Analysis in Data Mining

*Soma Roychowdhury*

## 1. Introduction

All fields of study gather and store data, e.g., retail companies gather information on sales, insurance companies gather information on claims, meteorological department measures and collects data related to weather conditions etc. with the purpose of making optimum decisions based on the information collected. The process of taking the raw data and converting them into meaningful information necessary to make decisions is the focus of this study. As we know, data mining is a process of exploring a large amount of data in order to find patterns and relationships among the variables and then validating the patterns and/or relationships obtained to a new subset of data. Data mining is based on conceptual principles of statistics including traditional 'Exploratory Data Analysis' (EDA) and Modelling (Han and Kamber, 2001). In this paper we will focus on various EDA tools.

## 2. Necessary Steps involved in an EDA Procedure

In an exploratory data analysis procedure, at first the problem should be defined before going into the details of the procedure. Once the problem is defined we can proceed for the data preparation, which is the most elaborate step in an EDA procedure. Further analysis can be done only after preparation of the data. Otherwise, our analysis may lead to a wrong decision and our entire attempt may go in vain. Next we implement the analysis followed by the deployment of the result. Thus we may divide the EDA procedure in the following four broad steps, the details of which are discussed in the subsequent sections :

- Problem definition
- Data preparation
- Data analysis
- Deployment of the results

### 2.1 Problem Definition

It is important to define the scientific problem to be solved. It is essential to understand how the problem will be addressed by the data analysis procedure. This step is very important in the sense that it gives a focus plan to be executed. It will ensure that all the important issues are taken into account.

### 2. 2 Data Preparation

In data mining literature, this step is known as initial exploration of data. Prior to starting any data analysis the data should be characterized, cleaned, transformed and partitioned into appropriate form for further processing. Preparation of data is one of the most time consuming parts of any data analysis or data mining exercise. How the data have been collected and prepared is critical to the confidence with which decisions can be made. The data should be reliable, and it should relate to the target population. The quality of the results obtained through data analysis depends on the quality of the data.

Data can be primary or secondary. Primary data can be obtained through :

- Direct observation method
- Designed experiment method
- Survey method (Interview method, mailed-questionnaire method)

Secondary data are compiled from published sources, like journals, government publications etc.

- *Direct observation method* : The units are observed and measured directly, as they appear in nature.
- *Designed experiment method* : By an experiment here they are obtained from treatment group and control group.
- *Survey method* : Collecting data through interviewing or mailing questionnaire.

### *2.2.1 Types of Databases*

The databases are generally of the following types:

- *Operational database* : This database contains on-going business transactions. They are accessed constantly and updated regularly.
- *Data warehouse/Data mart* : This contains cleaned, normalized data.
- *Historical database* : This database contains data of old business transactions, old experiments, surveys, polls etc.
- *Purchased data* : This database contains data purchased from other sources. This is necessary when in-house data are not sufficient.

### *2.2.2 Understanding Data-classification*

The procedure to be adopted to analyze a data set depends on the type of data. Major classification is as follows :

- Qualitative

  - *Ordinal*
  - *Non-ordinal*

- Quantitative

  - *Discrete*
  - *Continuous*

### 2.2.3 A Caution : Combining Data obtained from Different Sources

Sometimes, the data need be pulled together for the purpose of analysis, and pulling the data from different sources may introduce difficulties in interpreting them, if not pulled properly. When data from several sources or tables are combined, there is a possibility that some unreported variables (lurking variables) may cause a reversal of the findings.

Let us discuss a real-life example on kidney stone treatment, where the success rates of two treatments, A and B, are compared (Charig et al., 1986, Julious and Mullee, 1994).

The following table shows the number of cases cured and not cured and overall success rates under treatments A and B :

| | *Cured* | *Not cured* | *Total treated* |
|---|---|---|---|
| Treatment A | 273(0.78) | 77(0.22) | 350 |
| Treatment B | 289(0.83) | 61(0.13) | 350 |

This seems to show treatment B has a better success rate.

Now consider that the sizes of all stones are not equal. The size can be classified as either large or small, and the data can be split accordingly as follows :

| | *Treatment A* | | | *Treatment B* | | |
|---|---|---|---|---|---|---|
| | *Cured* | *Not cured* | *Total* | *Cured* | *Not cured* | *Total* |
| Large stone | 192(0.73)* | 71 | 263 | 55(0.69)* | 25 | 80 |
| Small stone | 81(0.93)* | 6 | 87 | 234(0.87)* | 36 | 270 |
| Total | 273(0.78)* | 77 | 350 | 289(0.83)* | 61 | 350 |

* *The proportions are shown in parentheses.*

*Conclusion is reversed.*

If we include data about kidney stone size, the same set of treatments reveals a different answer.

This may be explained as :

> The doctors tend to give the serious cases (large stones) the better treatment A, and the not-so-serious cases (small stones) the inferior treatment B. So the totals are dominated by the larger groups, the groups having the number of serious cases treated by treatment A and not-so-serious cases by treatment B.

We have to be very careful about combining data described in proportions remembering the fact $\frac{a}{b} > \frac{c}{d}, \frac{e}{f} > \frac{g}{h}$ does not necessarily imply $\frac{a+e}{b+f} > \frac{c+g}{d+h}$.
The reversal of conclusion when data are combined from several sources or groups is called reversal paradox or Simpson's paradox. It happens to the kidney stone data given here. The 'size of the stones' plays the role of lurking variable here.

### 2.2.4 Cleaning of data

It is important to clean the data by resolving any ambiguities, removing errors, redundancy and problematic data (Dasu and Johnson, 2003). Secondary data need more scrutiny, as it was initially collected with some objective, but later used for some other purposes. For variables measured in nominal and ordinal scales where there are fixed number of possible values, it is possible to scrutinize all possible values to uncover any mistakes or inconsistencies. Any assumptions made concerning the possible values that the variable can take should be tested. Cleaning of data on variables measured on interval or ratio scale needs more attention, since they can take any possible value within a range. It may happen that an observation in the data set is missing. Where there is knowledge and meaning for the missing data value, we may replace the value on the basis of the knowledge. There are methods to estimate the missing data value. Any duplicate entries should be removed.

### *Detecting an outlier*

It is important to detect outliers in the data set, if any. There could be many possible reasons for the presence of outliers in a data set. It could even be due to measurement error. An outlier may also be a genuine data point. There are several methods of detecting outliers, if any. One such method is as follows :

- Compute IQR $= Q_3 - Q_1$
- Obtain ($Q_1$–1.5 IQR, $Q_3$+1.5 IQR) and ($Q_1$–3IQR, $Q_3$+ 3 IQR)

Any observation falling beyond ( $Q_1$–1.5 IQR, $Q_3$ + 1.5 IQR) is identified as a suspect outlier, and any value falling outside ($Q_1$–3IQR, $Q_3$+ 3 IQR) is known as an extreme outlier.

If one has a reasonable idea about the standard deviation of the data set, then if the suspected point deviates from the mean by some predetermined multiple of it, it may be considered to be an outlier. To detect a possible outtier we compute M as given by

$$M = \frac{|suspect - mean|}{s.d.}.$$

If $M > 4$, then the suspect may be considered to be an *outlier*. This is called the *huge error rule*. If the standard deviation (s.d.) is known, then the probability of such decision could be as large as 0.999 or more.

Various graphs can also give us some idea about the presence of possible outliers. Histograms provide a clear way to view the data. It can give us the idea of the central values, the shape, the range of values, as well as the presence of any outliers. It helps identify the outliers by eye inspection. We look for data points those are outside the range of expected values in that case. Box plots also help in identifying the outliers. Box plots combine visual detection of outliers with a bit of statistics.

Another approach looks at how much each data point deviates from mean. The easiest way of doing this is to transform the data values into *z*-scores (given by $z = (x - \bar{x})$ / *s.d.*). Each data point will then represent how far it is from the mean, in s.d. units. The cut-off point between what is expected and what is an outlier is very case specific but usually any value having *z* score over 3.00 or under –3.00 is viewed with suspicion.

### *2.2.5 Data Transformation*

In order to apply some meaningful EDA procedure, it is necessary to apply certain mathematical transformations to the data. For example, to apply parametric tests we need the assumption that the data are normally distributed. Some tests make other demands on the data, which need to be satisfied. Common data transformation includes taking square roots, reciprocals and logarithms raising variables to positive integral powers. For data expressed in proportions, the logit transformation, $f(p) = p/(1-p)$, is often used. It should be noted that the common people may have hard time in making sense of the results derived using the transformed data in some cases, though it is statistically correct to transform the data.

We can consider the following guidelines regarding the transformation of data. Do not transform the data if

- the deviation from normality and linearity is not too extreme
- the sample size is large, say, over 30
- the statistics to be used whose robustness is known.

Now the question is which transformation to use. We can think of distributions as ranging from extremely skewed to the right to extremely skewed to the left. We can use the following transformation depending on the data types :

| *Shape* | *Transformation* |
|---|---|
| Reverse J shaped | $1/x$ |
| Severe skew to the right | $\log x$ |
| Moderate skew to the right | $\sqrt{x}$ |
| Moderate skew to the left | $-1/\sqrt{x}$ |
| Severe skew to the left | $-1/\log x$ |
| J-shaped | $-1/x$ |

- Normalization
  - (i) *z*-score : New value = (value – mean)/s.d.
  - (ii) Min-max normalization within (*a*,*b*) :

$$\text{New value} = a + (b - a)\frac{(\text{value} - \text{original min})}{(\text{original max} - \text{original min})}$$

  - (iii) Box-Cox transformation:

When a variable does not conform to the assumption of normality, but the data analysis method requires the data to come from a normal distribution, then one may use the following transformation, known as Box-Cox transformation, given by

$$\text{New value} = \left\{(\text{value})^{\lambda} - 1\right\} / \lambda, \lambda \succ 1.$$

With the same purpose the log-transformation is also used.

• *Value mapping or Coding*

Ordinal variables described by texts may be required to be converted to numbers. For example, the response of a patient to a therapy recorded as low, medium and high may be replaced by the codes 0, 1 and 2. A nominal variable can similarly be handled, e.g.,'0' may indicate the absence and '1' may indicate the presence of a category, like a subject is a male or a female, left-handed or not etc.

• *Discretization*

By converting continuous data into discrete values we may lose information. But these conversions are necessary in many situations. There are some techniques, which can process categorical data, and hence conversion of continuous data into discrete values makes the data usable to these methods. For example, a continuous variable 'credit score' may be categorized as poor, average, good, excellent. This type of conversion is called 'binning'.

## 2.3 Data Analysis

After preparing the data we can proceed for analyzing the data in order to make meaningful conclusions, and then we can deploy the result we obtain through data analysis. In data analysis step generally we have the following objectives :

- Summarizing data
- Finding hidden relationship
- Predicting unknown variables

*Summarizing data*

Summarization is a process where a data set is reduced for interpretation of data without sacrificing any important information. Any measures of central tendency dispersion, skewness, kurtosis can be referred to in this context.

*Finding hidden relationship*

Important facts, relationships, trends in the data which are not very obvious from a summary measure alone are required to be identified. To discover the hidden relationships we have to look at the data from various angles.

*Predicting unknown variables*

Prediction is the process where an estimate is obtained for some variable the value of which is unknown. For example, a retail company would like to predict, using historical data, the type of products that a specific group of consumers may be interested in.

There is a great deal of interplay among the above three objectives. It is important to summarize the data before making any prediction and finding hidden relationships. On the other hand, prediction or exploring the hidden relationships may help in summarizing the data towards making a decision.

### 2.3.1 *Tools and Techniques used in Summarizing Data, Finding Hidden Relationships and Making Predictions*

Following are some of the statistical tools used to achieve the objectives as described above :

- Summary tables

- Graphs
- Descriptive Statistics
- Inferential statistics
- Grouping
- Mathematical models

### *Summary tables*

The data are arranged and presented in the form of a table in order to make them more easily comparable, comprehensible and mobile.

### *Graphs*

Presenting data graphically allows the eyes to visually identify the patterns and relationships. Box plot, dot plot, stem-and-leaf plot, histogram etc. are most commonly used graphs. Unfortunately, in data mining, scatter plots are not always very useful. If there are too many data points, then the points in the scatter plot become so dense that we cannot get a good idea about the pattern of the data, as the nature of change will not be that obviously understood from the plot.

### *Descriptive statistics*

Different descriptive statistics can be used to describe the whole data set. Mention may be made of various measures of center, spread, asymmetry, association of variables etc..

### *Inferential statistics*

Different inferential statistics (point and interval estimators, test-statistics) can be used to make conclusions and verify the claims regarding the concerned parameters with certain level of confidence.

### *Grouping*

The methods for organizing the data into smaller groups (subsets) of related observations are called grouping. This is very important for exploratory data analysis and data mining exercise in order to find the hidden relationships among the variables that are not obvious from summary statistics. Different methods of grouping are :

(i) Grouping by values or ranges
(ii) Grouping by similarity measures or distance measures and clustering
(iii) Grouping by associative rules
(iv) Grouping by decision trees

#### (i) *Grouping by values or ranges*

While grouping the data set by values or ranges we create several queries. Each query will create subsets of observations. This set could then be examined to determine whether some interesting relationship exists among the groups. We may query the data set in variety of ways, e.g., in order to group the cars of different models according to the fuel efficiency of the cars we may query on the number of cylinders the cars have. Sometimes, we may require the answers to more complex queries in order to form the groups, such as the number of cylinders as well as the weights of the cars can form the basis of grouping.

#### (ii) *Grouping by similarity measures or distance measures and clustering*

Grouping of data may be done by observing the similarity among the observations. To determine the closeness or similarity between two observations, it is necessary to define precisely what we mean by 'similar'.

Let us define

S(i, j) = similarity measure between objects i and j
d(i, j) = dissimilarity measure between objects i and j
S(i, j) and d( i, j) may be related in some way, like

$$d(i, j) = 1 - S(i, j)$$

$$\text{or} \quad d(i, j) = \sqrt{\{2\,(1 - S(i, j))\}}.$$

The term 'distance' is used to refer to a dissimilarity measure derived from the characteristics describing the objects. A metric is a dissimilarity measure that satisfies the following three conditions :

1. $d(i, j) \geq 0, \forall\, i,\ j,\ i \neq j$ and $d(i, j) = 0$ if $i = j$
2. $d(i, j) = d(j, i)\ \ \forall\ i,\ j$
3. $d(i, j) \leq d(i, k) + d(k, j)\ \ \forall\ i, j, k$ (Triangle inequality).

*Different distance measures*

- Euclidean distance between two p-dimensional observations A and B, is given by

$$d(A, B) = \sqrt{\sum_{k=1}^{p} (x_k - y_k)^2}$$

where $A = (x_1, x_2, ..., x_p)$ and $B = (y_1, y_2, ..., y_p)$.

- Weighted Euclidean distance between two p-dimensional observations A and B is given by

$$d(A, B) = \sqrt{\sum_{k=1}^{p} w_k (x_k - y_k)^2}\ ,$$

where $w_1, w_2, \ldots w_p$ are respective weights.

- Minkowski's distance or $L_\lambda$-metric between two p-dimensional observations A and B,

$$d(A, B) = \left\{ \sum_{k=1}^{p} | x_k - y_k |^\lambda \right\}^{1/\lambda}, \lambda \geq 1.$$

Note that the Euclidean distance is a special case of Minkowski's distance. It can be seen as $L_2$–metric. For $\lambda = 1$, Minkowski's distance is known as Manhattan distance or city-block distance and is given by

$$d(A, B) = \sum_{k=1}^{p} | x_k - y_k |.$$

For $\lambda \to \infty$, we define $L_\infty$-metric as

$$d(A, B) = max_k | x_k - y_k |.$$

The following example gives an idea as to how the measures can be used to evaluate the similarity between the observations. Consider the following three 5-dimensional observations on A, B and C (i.e., five variables measured on three subjects) :

| | Variable 1 | Variable 2 | Variable 3 | Variable 4 | Variable 5 |
|---|---|---|---|---|---|
| A | 7.5 | 8.2 | 4.3 | 5.7 | 2.8 |
| B | 6.5 | 7.6 | 5.2 | 5.7 | 3.1 |
| C | 9.5 | 7.9 | 6.5 | 6.8 | 5.2 |

The Euclidean distances between the pairs of observations are determined below :

$$d(A, B) = \{(7.5 - 6.5)^2 + (8.2 - 7.6)^2 + (4.3 - 5.2)^2 + (5.7 - 5.7)^2 + (2.8 - 3.1)^2\}^{1/2}$$
$$= 1.5033$$
$$d(A, C) = 3.9875$$
$$d(B, C) = 4.0497.$$

Thus B is more close to A than to C, A is more close to B than to C, and hence A and B are more similar compared to B and C or A and C here. While grouping, it is more rational to put A and B in the same group. Likewise the other distance measures can also be determined for the purpose of grouping.

For multivariate binary data we can count the number of variables on which two subjects take the same or different values and define the similarity measure, matching coefficient or similarity coefficient, as

$$\frac{f_{11}+f_{00}}{f_{11}+f_{10}+f_{01}+f_{00}},$$

where $f_{11}+f_{10}+f_{01}+f_{00}=\text{p}$, the total number of binary variables and $f_{11}, f_{10}, f_{01}, f_{00}$ are given in the following table :

| | $j=1$ | $j=0$ |
|---|---|---|
| $i=1$ | $f_{11}$ | $f_{10}$ |
| $i=0$ | $f_{01}$ | $f_{00}$ |

Sometimes, it is inappropriate to include cell (0,0) (or (1,1)), where 0 (or 1) indicates the presence of a very common quality on which comparison of similarity is meaningless.

Then a modification of the similarity coefficient can be obtained which is known as Jaccard coefficient. If '0' indicates the absence of a rare quality, then Jaccard coefficient is defined as

$$\frac{f_{11}}{f_{11}+f_{10}+f_{01}},$$

and hence the Jaccard distance is given by (1 – Jaccard coefficient), i.e.,

$$\frac{f_{10} + f_{01}}{f_{11} + f_{10} + f_{01}}.$$

The following example illustrates how the similarity coefficients can be obtained from a binary data set :

| | *Variable 1* | *Variable 2* | *Variable 3* | *Variable 4* | *Variable 5* |
|---|---|---|---|---|---|
| A | 1 | 1 | 1 | 0 | 1 |
| B | 0 | 1 | 0 | 1 | 0 |
| C | 1 | 1 | 0 | 0 | 1 |

Jaccard coefficient of similarity between A and B, J(A,B) = 1/5.

Similarly, J(B,C) = 1/4, J(C, A) = 3/4. This indicates that the similarity between A and C is maximum of all three pairs of subjects. So, A and C can be grouped together.

## Clustering

Grouping of subjects into clusters, the sets of related subjects, on the basis of similarity or distance measures is the basic objective of clustering. It is an unsupervised method of grouping. The subjects within each group or cluster are more similar to the other subjects within the group than to the subjects within any other group. The various methods of clustering can be classified as follows :

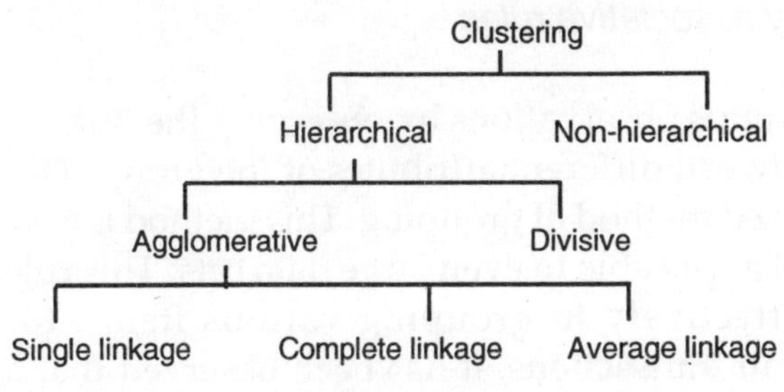

A hierarchical agglomerative clustering uses a bottom-up approach of clustering as it starts with each subject as a member of a separate cluster and progressively merges clusters together until all subjects are a member of a final single cluster. A non-hierarchical clustering (*k*-means clustering) uses a top-down approach of clustering as it starts with a predefined number of clusters and assigns subjects to those clusters. It can handle a very large data set efficiently, whereas the hierarchical clustering can handle data sets of size less than 10,000. But the non-hierarchical clustering might fail to give an optimal grouping if the data set contains many outliers. In agglomerative hierarchical clustering, the distance between a single observation and a cluster is determined on the basis of a pre-defined linkage rule. In single linkage method the clusters are formed by minimizing the minimum distance between a group of observations and a new observation. In complete linkage method, the maximum distance is minimized, whereas in average linkage method, the average distance is minimized.

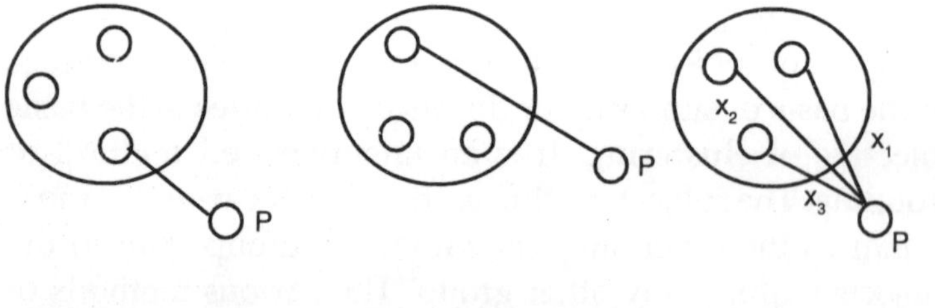


Single linkage rule　　Complete linkage rule　　Average linkage rule

(iii) *Grouping by associative rules*

This method groups observations by observing the links or associations between different attributes of the group. This is an unsupervised method of grouping. This method is easy to interpret and applicable to even large data sets. This rule can be used effectively in grouping various items, for example, in retail transactions, it has been observed that a customer who buys bread, often buys butter, jam or jelly. So, these items can be put together, e.g., corn flakes should be placed in the shelves close to milk in a grocery.

A retailer can make a package of pen, pen-stand together with pencil, eraser, pencil-sharpener, pencil box for the school or college-goers, toys with batteries, digital cameras with camera bag, memory stick or card, battery, tripod etc. with the same consideration.

### (iv) *Grouping by decision trees*

It is a supervised method of decision-making, because it utilizes the knowledge from the response data to form the groups. Here grouping is made using the decision trees that are made up of a series of decision points, where the set of observations is split based on some criteria. This method attempts to group the subjects based on the responses received from a number of questions that are asked where the subsequent questions are also based on the response to the previous question. The data set is initially divided into two or more mutually exclusive and exhaustive groups and subsequently they are split to various branches based on some criteria.

### *Mathematical models*

Mathematical models are equations or processes that express the response variable in terms of the descriptors. Different models for prediction are used in many situations where an estimate or forecast is required. Such models usually estimate the response variable on the basis of covariates called descriptors or inputs. The quality of the model depends on how well the model is able to predict the output variable correctly for a given set of input variables or descriptors. Models based on a single input variable are called simple, whereas those based on two or more input variables are called multivariate models. Correlation analysis as well as other statistical techniques can be used to identify which input variables appear to be the most influential for prediction purpose. Predictive models can be optimized by fine-tuning different parameters appearing in the model.

When the response is a categorical variable, the predictive model is called a classification model, and when the response is a continuous variable, the model is called a regression model. A predictive model is built using the training data where the data for both of the response and input variables are available. Once the model has been built, only the data on input variables are fed into the model to predict the response variable. Sometimes, for building up of a model the training data set is divided into two sets – one is used to build the model and other, to assess the quality of the model built. Classification model is built to assign observations into two or more distinct categories. Classification is a supervised method of prediction, since it utilizes the class labels of each training sample to build the classification rules. A multiple regression model is used as a prediction model where the continuous response variable depends on more than one input variable. Neural network is a prediction model based on a series of input descriptor variables connected to output response variables where each connection has a weight associated with it. It utilizes a training set to generalize the relationship between the input variables and output variables. Input descriptors (also called input nodes) and output responses (also called output nodes) are connected through some hidden layer of nodes. Normally, the number of hidden layers varies from 0 to 5. Neural networks are used to derive the complex linear and non-linear relationships, for categorical and continuous data. Neural networks are less sensitive to noise than the regression models.

## 2.4 Deployment of the Results

At this step of EDA the results or solutions obtained through the analysis of data is translated into a benefit to the business and society. It measures and monitors the solution's performance.

In data mining, the conclusions drawn on the basis of EDA techniques always bear some risk of wrong decision. This chance of wrong decision is present in all the tools of decision-making under uncertainty. While estimating the model parameters or predicting some output variable, correct specification of the loss function is very important. It has been noticed that the use of symmetric loss functions is predominant in the studies related to the estimation and prediction problems. Although these forms of loss functions are mathematically more easily tractable, but they are not enough capable of capturing the true loss structure. The shape of the relevant loss function may be asymmetric in contrast to the researchers' favourite symmetric loss function, such as the squared error loss. Mention may be made of some estimation and prediction problems where the impact of loss can be different for equal amount of over-estimation and under-estimation, and for equal amount of positive and negative prediction errors, i.e., there exists a natural imbalance in the economic results of estimation and prediction errors of same magnitude and of opposite signs. In those situations the use of the symmetric loss functions that are widely used in the context of efficient estimation and prediction, does not give a very good result. In particular, let us consider the following examples. In agricultural experiment, the under-use of fertilizer or spray of pesticides may result in a low yield, but an equal amount of over-use of fertilizer or pesticide may damage the whole crop. Similar cases can be found in other fields too. Consider another decision problem relating to whether the possibly polluted environment (or site) needs clean up based on the magnitude of the pollutant, duration of exposure to pollutant etc. The usual decision rule based on the UCL (95% Upper Confidence Limit of the average toxic concentration) is too conservative and suffers from a large amount of over-estimation of the true average toxic concentration level, which in turn, may result in unnecessary costs for cleaning the site or collecting more samples from the site. An asymmetric loss function may be very useful in the above context, which will assign different losses for false positive decision due to over-estimation (i.e., unnecessary remediation) and false negative decision due to under-estimation (i.e., no remediation for a polluted site which actually needs a clean up).

## References

Charig, C. R., Webb, D.R., Payne, S. R. and Wickham, O.E. (1986), Comparison of Treatment of Renal Calculi by Operative Surgery, Percutaneous Nephrolithotomy and Extracorporeal Shock Wave Lithotripsy, *Br. Med. J.* (*Clin. Res. Ed.*), 292 (6524), 879-882.

Dasu, T. and Johnson, T. (2003), *Exploratory Data Mining and Data Cleaning*, John Wiley and Sons.

Han, J. and Kamber, M. (2001), *Data Mining : Concepts and Techniques*, Academic Press.

Julious, S.A. and Mullee, M.A. (1994), Confounding and Simpson's Paradox, *Br. Med. J.*, 309 (6967), 1480-1481.

# 5

# Non-parametric Statistical Methods in Medical and Social Science Research

*K.R. Sundaram*

## 1. Introduction

Sometimes it is difficult to describe an appropriate theoretical model for the experiment. The real difficulty often comes after the model has been defined in the form of finding the probabilities associated with the model. Many reasonable models have been invented for which probability solutions have been found. This body of Statistics, i.e., applying the probability model for making inferences from the sample of experiment in order to arrive at valid conclusion is known as 'Parametric' statistical methods.

The commonly applied inference methods such as Student's t-test and Fisher's F-test, belong to this category. In parametric method, exact solutions for the approximately suitable probability model are found. However, in the late 1930s, a different approach to the problem of finding probability began to gather momentum. This approach involves making few changes in the model and using simple unsophisticated methods to find out the desired probability. Thus, approximate solutions to the exact problems were found as opposed to the exact solutions to approximate problems. These new packages of Statistical Methods became to be known as 'Non-parametric' methods.

The advantages of non-parametric statistics over parametric statistics are as follows :

1. Simple model 2. Easy computability 3. No assumption on the form of population distribution of the variable 4. No need of larger sample for making inferences.

In case of applying parametric model, the specific form of distribution of the variable in the population is required. Also, the computability is sometimes not easier and hence not quicker. However, randomness of the sample is required in applying non-parametric methods as in case of parametric methods. Since the assumption of specific form of distribution of the variable is not required, the non-parametric methods are also sometimes called 'Distribution free methods'.

## 2. Non-parametric Inference

### 2.1 Statistical Estimation

| | | Parametric | Nonparametric |
|---|---|---|---|
| 1. | Representative value | Mean, Median Mode | Median, Mode |
| 2. | Variation | Standard Deviation (SD) | Quartile Deviation, Range |
| 3. | Correlation | Person's Product Moment Correlation Coefficient ( $\rho$ ) | Spearman's Rank Corr. Coefficient ( $\rho_R$ ) |
| 4. | Intervals for the estimate | Mean ± SD | Quartiles (Q1, Q3), Percentiles (P3, P97), (P5, P95), (P10, P90) |

#### 2.1.1 Spearman's Rank Correlation

In parametric estimation, correlation coefficient between two continuous variables is obtained by computing Pearson's p. The corresponding non-parametric correlation coefficient is obtained by computing Spearman's correlation coefficient $\rho_R$.

$$\rho_r = 1 - \frac{6\sum d_i^2}{n(n^2-1)},$$

where $d_i$'s are the differences between the ranks of the values of the two variables or between the values of the same variable by two methods.

*Examples*

1. Ranking of 10 children by 2 educationalists on I.Q.

| Sl. No. | 1 | 2 | 3 | 4 | 5 | 6 | 7 | 8 | 9 | 10 |
|---|---|---|---|---|---|---|---|---|---|---|
| A | 4 | 10 | 3 | 1 | 9 | 2 | 6 | 7 | 8 | 5 |
| B | 5 | 8 | 6 | 2 | 10 | 3 | 9 | 4 | 7 | 1 |
| di | -1 | 2 | -3 | -1 | -1 | -1 | -3 | 3 | 1 | 4 |

$$\rho_R = 1 - \frac{6\sum d_i^2}{n(n^2-1)} = 1 - \frac{6 \times 52}{10 \times 99} = 0.685$$

2. Age (years) and systolic blood pressure (SBP) of 6 men are given below :

| Age : | 25 | 30 | 36 | 45 | 65 | 70 |
|---|---|---|---|---|---|---|
| SBP : | 120 | 125 | 136 | 140 | 135 | 150 |

$\rho$(P) = 0.8198, $\rho_R$(NP) = 0.8286 (P: Parametric, NP: Non-parametric)

3. Weight (kgs.) and cholesterol values (mg/dl.) of 6 men are given below :

| Age : | 50 | 80 | 65 | 90 | 60 | 75 |
|---|---|---|---|---|---|---|
| Cholesterol : | 180 | 450 | 350 | 650 | 400 | 500 |

$\rho$(P) = 0.554, $\rho_R$ (NP) = 0.6.

### *2.1.2 Location Parameters : Quartiles and Percentiles*

$Q_1$ = (¼) $(n+1)^{st}$ value, $Q_2$ = (½)$(n+1)^{st}$ value, $Q_3$ = (3/4) $(n+1)^{st}$ value of all the values in the sample arranged in ascending order. $Q_1$ is called the first quartile or lower quartile. Twenty five per cent of the values in the sample will lie below $Q_1$ and 75% above. Similarly, $Q_3$ is called the $3^{rd}$ quartile or upper quartile. Seventy five per cent values in the sample will be below $Q_3$ and 25% values will be above $Q_3$. $Q_2$ is called the median value. Fifty per cent values will be below $Q_2$ and 50% will be above $Q_2$.

$P_5$ is called the 5th percentile. Five per cent values will be below $P_5$ and 95% will be above. $P_5 = \{5(n+1)/100\}^{th}$ value. Similarly, the other percentiles like $P_3$, $P_{10}$, $P_{90}$, $P_{97}$ can be defined and computed. Note that

$$P_{25} = Q_1,\ P_{50} = Q_2 = \text{median},\ P_{75} = Q_3.$$

In smaller samples the location parameter values may not be very accurate. When the sample size goes on increasing the values will be more accurate.

The following data give age, arranged in order, of 15 patients admitted in a Cardiac unit of a hospital :

45, 47, 52, 55, 60, 62, 62, 63, 64, 66, 69, 70, 72, 72, 74, 80
$Q_1 = 4^{th}$ value = 55, $Q_3 = 12^{th}$ value = 72, $Q_2$ = median = $8^{th}$ value =64,
$P_3 = \{(3 \times 16)/100\}^{th}$ value = $0.48^{th}$ value=$1^{st}$ value=45,
$P_{97} = (0.97 \times 16)^{th}$ value = $15.52^{th}$ value=$15^{th}$ value = 80
and so on.

As mentioned earlier, the position of the location parameters will be better defined and the values will be more accurate in larger samples.

### 2.1.3 Non parametric Linear Regression

The linear regression equation is Y = a + b X, where b is the regression coefficient or slope, and a is the Y-intercept.

b is obtained as the value of the median of $S_{ij}$ -values, where

$$S_{ij} = (Y_j - Y_i) / (X_j - X_i),$$

$i \neq j$, i, j = 1, 2,..., n being the sample size, and (X,Y)'s are the sample pairs of values. If n = 10, there will be $^{10}C_2 = 45$ $S_{ij}$-

values. After arranging them in ascending order, b will be the median of the 45 $S_{ij}$-values. This is called Theil's slope estimator.

The value of a is obtained as follows :

> a = median of the n (n+1)/2 pairwise averages of the terms $Y_i - (bx_i)$.
>
> For the above example there will be 55 pairwise averages of $Y_i$–($bx_i$) terms. This method is due to Dietz.

Theil's and Dietz methods of computing a and b values are explained in the following example. For simplicity, the method is explained for a sample of size 5. The following data give the age and systolic blood pressure of 5 males :

| Age : | 45 | 56 | 60 | 70 | 55 |
|---|---|---|---|---|---|
| SBP : | 130 | 135 | 150 | 140 | 150 |

The number of values of $(Y_j - Y_i) / (X_j - X_i)$ will be $5C_2$ or 10. They are :

(45-56)/(130-135) = 2.2, (45-60)/(130-150)=0.75 and so on. Thus the 10 values obtained by doing this operation are :

2.2, 0.75, 2.5, 0.5, 0.27, 2.8, -0.067, -1.0, infinity, -1.5.

These values, arranged in ascending order, are :

-1.5, -1, -0.067, 0.27, 0.5, 0.75, 2.2, 2.5, 2.8 and infinity.

The median of these values is the average of 5th and 6th values, i.e., (0.5 + 0.75)/2 = 0.625. Hence b = 0.625.

Now $T_1 = Y_1 - bX_1 = 130 - (0.625 \times 45) = 101.875$.

Similarly, the value of $T_2$, $T_3$, $T_4$ and $T_5$ are computed. Thus five T values are :

101.875, 100, 112.5, 96.25 and 115.625.

Then $R_1 = (T_1+T_1)/2 = 101.875$, $R_2 = (T_1+T_2)/2 = (101.875+100)/2 = 100.9375$ and so on.

There will be $(5\times6)/2 = 15$ R-values for a sample of size 5. The fifteen R values are :

101.875, 100.9375, 107.1875, 99.0625, 108.75, 100, 106.25, 98.125, 107.8125, 112.5, 104.375, 114.0625, 96.25, 105.9375, 115.625.

The median of these R values is the value occupying the $8^{th}$ position, when these values are arranged in ascending order.

Hence the median = 105.9375 = a.

Thus the linear regression equation is Y = a +b X, i.e.,

$$Y = 105.9375 + 0.625\ X.$$

The corresponding linear regression equation obtained by applying parametric least squares method on the same data set is: Y = 117.546 + 0.41 X.

### *2.1.4 Non-parametric Confidence Limits for the Median*

A 95% confidence limits for the median are given by the values in the ascending order in the sample occupying the positions A and B, where

$$A = np - 1.96\ \{np(1-p)\}^{1/2}$$

and $B = np + 1.96\ \{np(1-p)\}^{1/2}$,

n is the sample size and p = 0.5.

For example, the number of cases of malaria reported in a village during the last 12 months, arranged in ascending order are :

10, 20, 35, 40, 60, 78, 90, 120, 132, 145, 150, 160.

Here median = average of $6^{th}$ and $7^{th}$ values = (78+90)/2 = 84.

n =12, p=0.5, A = (12 ×0.5) – $\{12\times0.5\times0.5\}^{1/2}$ = 6 –1.732 = 4.268 B = 6+1.732 = 7.732.

The values of A and B should be rounded off to the next integer to find the actual positions, i.e., A = 5 and B = 8, i.e., the values occupying $5^{th}$ and $8^{th}$ positions, i.e., 60 and 120 will give the confidence limits. Since the actual positions are obtained by rounding off to the next integer, the confidence limits will not exactly be 95%. It will be slightly more or less than 95%.

Another method for computing the confidence limits for median is :

LL = value occupying the position A,

where A = (n/2) –1.96 {√n/2} and

UL = value occupying the position B,

where B = n+1–A, LL and UL denote the lower and upper limits, respectively.

This method is suggested when n is greater than 100.

For the example given above,

A = 6 – 1.96×(3.464/2) = 2.61=3, B = 13–3 = 10,

Hence LL = $3^{rd}$ value = 35 and UL =$10^{th}$ value =145.

When n is smaller than 100, the following method is suggested :

Refer to the Table A for the values of n, taken vertically

and at the chosen level of significance. For example given above n = 12 and for 95% confidence limits, A = 3 and B = n+1–3 =10. Hence the 95% confidence limits are those values occupying the 3$^{rd}$ and 10$^{th}$ positions, i.e., 35 and 145.

This method is used to calculate the confidence limits for the median in case of comparatively large sample size, say more than 20 or so. The assumption is based on normal approximation of the distribution of the values of the variable. In case of smaller samples, actual limits have to be found by using a Statistical Table and the procedure is a bit laborious. Larger and larger the sample size, the normal approximation will be more valid and the confidence limits will be more accurate.

It has been shown by some researchers that the power of many parametric methods is lesser compared to the corresponding parametric methods. Hence, it is suggested that one should try his/her best to apply the parametric inference methods if the conditions for applying such methods are met. This can be achieved by suitable transformation of the values of the variables. If all these approaches fail, then the only method of arriving at conclusions with some validity and robustness is by applying the non-parametric methods. Some of the commonly applied non-parametric tests of statistical significance are listed in the next section.

## 3. Non-parametric Test Procedure

Here we will discuss the following test procedures which are widely used in practice :

1. *Wilcoxon's rank sum test* : This is used for testing whether the two independent samples with respect to a variable come from the same population or not. This would be equivalent to see whether 'does one population tend to yield larger values than the other

population or do the two medians are equal or not.' This test corresponds to the normal test (Z) or the student's "t" test for two independent samples.

2. *Wilcoxon's signed rank test* : This is used for testing whether the differences observed in the values of the variable between two correlated populations (before and after design) are statistically different or not. This corresponds to the paired 't' test in parametric methods.
3. *Kruskal-Wallis One-way Analysis of variance* : This is used for testing whether several independent samples come from the same population or not. This corresponds to one-way analysis of variance in parametric method.
4. *Friedman's Two-way Analysis of variance* : This is used for testing whether the differences observed in the values of the variable between different time periods are statistically significant or not. This test corresponds to the two-way analysis of variance in parametric methods.

The above methods can be applied by ranking the observations appropriately and doing simple computations. Though, all these methods can be applied manually. Computer packages are available for applying these methods. The most commonly used computer package for non-parametric methods is MICROSTAT. Many other computer packages like BMDP, SPSS, SAS and SYSTAT also contain software programes for non-parametric methods. In case of comparatively large sample size, sometimes the ranking of observations, especially, when there are more than two groups, may be difficult and confusing. In such a situation, sorting the problems using computer packages may be easier and quicker.

Now we will give a brief description on the use of non-parametric test procedures and its parametric counterpart.

| | | | Parametric | Nonparametric |
|---|---|---|---|---|
| 1. | Comparison between two independent populations | Continuous | z-test<br>t-test | Wilcoxon's rank Sum test |
| | | Discrete | — | $x^2$-test |
| 2. | Comparison between correlated populations | Continuous | Paired 't' test | Wilcoxon's signed rank test |
| | | Discrete | — | McNemar's $x^2$-test |
| 3. | Comparison among several independent populations | Continuous | One-way Anova | Kruskal-Wallis One-way Anova |
| | | Discrete | — | $x^2$-test |
| 4. | Comparison among several correlated populations | Continuous | Two-way Anova | Freidman's Two-way Anova |
| | | Discrete | — | McNemar's $x^2$-test |

## *Example*

### (a) *Two Independent samples*

Intelligent quotient (IQ) of 5 normally nourished children (NN) and 4 malnourished children (MN), aged 4 years, are given below:

NN : 60, 80, 120, 130, 100
MN : 50, 60, 100, 45

Null hypothesis, $H_0$: The median IQs in the two groups are the same.

Alternate hypothesis, $H_1$: The median IQs in the two groups are different.

$n_1$ : sample size of the smaller group(MN) = 4
$n_2$ : sample size of the larger group(NN) = 5.

Rank all the values in the two groups taken together. Tied values should be given average ranks. For example, 60 is repeated twice. Hence, after giving rank 1 to the smallest value 45 and 2 to 50, the next highest value is 60, which is repeated twice, average of the ranks 3 and 4 is taken and a rank of 3.5 is given to each of these two values of 60. This method is repeated whenever there are tied values.

Ranks : NN : 3.5, 5, 8, 9, 6.5 MN : 2, 3.5, 6.5, 1

$T_1$ = sum of the ranks of smaller group (MN) = 13

$T_2 = S_1(n_1 + n_2 + 1) - 13$

$= 4(4 + 5 + 1) - 13 = 27$

T = smaller of $T_1$ and $T_2$ = 13

Assuming that the ranks are randomly distributed in the two groups, it is expected that the average ranks for each of the groups are approximately equal.

$$Z = \{|m - T| - 0.5\}/SD,$$

where m = mean of sum of the ranks = mean of $T_1$ and $T_2$

$= \{n_1(n_1 + n_2 + 1)\}/2$

$= \{4(4 + 5 + 1)\}/2 = 20,$

SD = standard deviation of m

$= \{(n_1 \times n_2)(n_1 + n_2 + 1)/12\}^{½}$

$= \{(4 \times 5)(4 + 5 + 1)/12\}^{½} = (200/12)^{½} = 4.1.$

The observed value of Z is

$$z = \{|20 - 13| - 0.5\}/4.1 = 6.5/4.1 = 1.6.$$

If z is less than 1.96, $H_0$ is not rejected. If $z > 1.96$, $H_0$ is rejected at 5% level of significance. Then $p < 0.05$.

If $z > 2.58$, then $p < 0.01$, and $H_0$ is rejected at 1% level.

If $z > 3.29$, $p < 0.001$, and $H_0$ is rejected at 1% level.

In the above example, $z < 1.96$, i.e., $H_0$ is not rejected.

The hypothesis of identical median IQ-values in the two groups is not rejected, i.e., the difference in median IQs between the two groups is not statistically significant.

Since the sample sizes in the two groups are very small, normal approximation may not be valid. Hence, p-value has been found by referring to the statistical Table $B_1$. For $n_1 = 4$ and $n_2 = 5$, at p = 0.05, the critical ratio (CR) = 11. Under $H_0$, T should be lesser than the CR. In the example, T = 13 >

11(CR), i.e., the difference in median IQ's between the two groups is statistically significant.

(b) *Dependent samples : Wilcoxon's signed rank test*

IQ values of 8 malnourished children of 4 years of age, before and after giving some nutritious diet for three months are given below:

| | | | | | | | | |
|---|---|---|---|---|---|---|---|---|
| Before (*b*) | 40 | 60 | 55 | 65 | 43 | 70 | 80 | 60 |
| After (*a*) | 50 | 80 | 50 | 70 | 40 | 60 | 90 | 85 |
| Difference(*d*) | 10 | 20 | -5 | 5 | -3 | -10 | 10 | 25 |
| Rank | 5 | 7 | -2.5 | 2.5 | -1 | -5 | 5 | 8 |

where d= $a - b$.

Rank the differences from smallest, without considering the sign of the difference. After giving the ranks, the corresponding sign should be attached.

$T+$ = sum of the ranks of the positive signs = 27.5
$T^-$ = sum of the ranks of the negative signs = 8.5.

Under $H_0$, it is expected that the sum of the ranks of the positive differences will be equal to the sum of the ranks of the negative differences. Assuming normal distribution for the differences, the test statistic is

$$Z = \{|m - T| - 0.5\} / SD,$$

where T = smaller of T+ and $T^-$ = 8.5

m = mean of sum of ranks = $\{n(n+1)\}/4 = \{8(8+1)\}/4 = 18$.
SD = $\{n(n+1)((2n+1)/24\}^{1/2}$
$= \{8(8+1)(16+1)/24\}^{1/2} = 7.14$.

Observed $Z = \{|8.5 - 18| - 0.5\} / 7.14 = 1.26 < 1.96$ ($p > 0.05$),

i.e., the difference in the sum of the positive ranks and the sum of the negative ranks is not statistically significant.

Increase in the IQ values after giving the nutritious diets for three months is not statistically significant.

For small sample sizes, Table C, should be referred to. For n = 8, the critical ratio at p = 0.05 is 2. Under $H_0$, T should be lesser than CR, but, in the example, T (=8.5) > CR(=2), i.e., the null hypothesis of no difference in the ranks before and after intervention is rejeced.

(c) *Independent samples* (*more than two groups*)

Intelligent quotient (IQ) of 5 normally nourished children (NN), 4 moderately malnourished children (MMN) and 5 severely malnourished children (SMN), aged 4 years, are given below :

NN : 60, 80, 120, 130, 100
MMN : 50, 60, 100, 45
SMN : 50, 40, 60, 35, 65

Kruskal-Wallis ANOVA (one -way ANOVA) :
Rank the values taking all three groups together :

| | | | | | | Total of ranks (R) | $R^2$ |
|---|---|---|---|---|---|---|---|
| NN (A) | 60 | 80 | 120 | 130 | 100 | | |
| Ranks | 7 | 10 | 13 | 14 | 11.5 | 55.5 | 3080.25 |
| MMN (B) | 50 | 60 | 100 | 45 | | | |
| Ranks | 4.5 | 7 | 11.5 | 3 | | 26.0 | 676.00 |
| SMN (C) | 50 | 40 | 60 | 35 | 65 | | |
| Ranks | 4.5 | 2 | 7 | 1 | 9 | 23.5 | 552.25 |
| Total | | | | | | 105 | 4308.50 |

Test statistic, $T = \frac{12}{n(n+1)} \sum \frac{R_i^2}{n_i} - 3(n+1),.$

Where $n = \sum_i n_i$

$T$ is distributed as $\chi^2$ distribution with (k-1) degrees of freedom (d.f.), where k is the number of groups.

In the example, there are three groups and if the average of two groups is available and the total of all values is also available, then the average of the third group can be obtained by arithmetic calculations.

For our example,

$$T = 0.0571\left[\frac{(55.5)^2}{5}+\frac{(26.0)^2}{4}+\frac{(23.5)^2}{5}\right]-45$$

$$= (0.0571 \times 895.55) - 45 = 51.17 - 45$$

$$= 6.17$$

$> 5.99$ ($\chi^2$ critical value for 2 d.f. at 5% level of significance, from Table D),

i.e., the differences in the median IQs among the three groups are statistically significant ($p < 0.05$).

p-value (for parametric one-way ANOVA) : 0.1952
p-value (for non-parametric one-way ANOVA): 0.0438.

The difference in the p-values obtained by applying parametric and non-parametric ANOVAs may be noted. Hence, it can be seen that parametric ANOVA should not be applied if the assumptions for applying parametric ANOVA are not satisfied.

ANOVA shows only whether the difference in median values among the three groups is statistically significant or not. To identify the pairs of groups (A and B, B and C, and A and C), which are significantly different, a *multiple range test* should be applied. The method is explained below :

- Compute the absolute differences : $|(R_1/n_1)-(R_2/n_2)|$, $|(R_2/n_2)-(R_3/n_3)|$ and $|(R_1/n_1)-(R_3/n_3)|$.
- Compute $w = \sqrt{\left[S^2\left(\frac{n-1-T}{n-k}\right)\right]}$, where $S^2 = \frac{n(n+1)}{12}$.
- Find $t$ from the Table E, at p=0.05 for (n-k) d.f.

- Compute $w_1 = tw\sqrt{\left(\frac{1}{n_1}+\frac{1}{n_2}\right)}$

$$w_2 = tw\sqrt{\left(\frac{1}{n_2}+\frac{1}{n_3}\right)}$$

$$w_3 = tw\sqrt{\left(\frac{1}{n_1}+\frac{1}{n_3}\right)}.$$

If $\left|\frac{R_1}{n_1}-\frac{R_2}{n_2}\right| > w_1$, the difference in the median ranks between NN and MMN is statistically significant ($p < 0.05$); otherwise the difference is not statistically significant.
Similarly we can make our conclusion for the other pairs if

$$\left|\frac{R_2}{n_2}-\frac{R_3}{n_3}\right| > w_2 \text{ and } \left|\frac{R_1}{n_1}-\frac{R_3}{n_3}\right| > w_3.$$

For the example, $S^2 = 17.5$.

$$w = \sqrt{\frac{17.5 \times (13 - 6.17)}{11}} = \sqrt{10.866} = 3.3$$

$$w_1 = 1.96 \times 3.3 \times \sqrt{\frac{1}{5}+\frac{1}{4}} = 4.34$$

$$w_2 = 1.96 \times 3.3 \times \sqrt{\frac{1}{4}+\frac{1}{5}} = 4.34$$

$$w_3 = 1.96 \times 3.3 \times \sqrt{\frac{1}{5}+\frac{1}{5}} = 4.09$$

$$\left|\frac{R_1}{n_1}-\frac{R_2}{n_2}\right| = \frac{55.5}{5}-\frac{26.0}{4} = 4.6 > 4.34\,(=w_1)$$

$$\left|\frac{R_1}{n_1}-\frac{R_3}{n_3}\right| = \frac{55.5}{5}-\frac{23.5}{5} = 4.6 > 4.09\,(=w_3)$$

$$\left|\frac{R_1}{n_2}-\frac{R_3}{n_3}\right| = \frac{56.5}{4}-\frac{23.5}{5} = 1.8 > 4.34\,(= w2)$$

i.e., the difference in median ranks of IQs between NN & MMN and between NN & SMN are statistically significant ($p < 0.05$) and the difference between MMN & SMN is not statistically significant ($p > 0.05$).

The table below gives the *p*-value from Student's *t*-test, Wilcoxon's Rank Sum test and multiple range tests (P and NP).

| *Group* | *t* | *WRS* | *MRP* | *MRNP* |
|---|---|---|---|---|
| NN & MMN | 0.102 | 0.0838 | <0.05 | <0.05 |
| NN & SMN | 0.009 | 0.0212 | <0.05 | <0.05 |
| MMN & SMN | 0.316 | 0.4587 | >0.05 | >0.05 |

The above given p-values show that the comparison of the groups applying student's 't' test between two groups, applying Wilcoxon's rank sum test and applying parametric multiple range test are not correct when there are more than two groups and if the assumptions for applying parametric test are not satisfied. The correct test in this situation will be the multiple range test for non-parametric ANOVA. The identical results above for MRP and MRNP are just by chance.

### (d) Paired (repeated) samples (more than two occasions): Friedman's Two-way ANOVA

IQ values of 8 malnourished children of 4 years of age, before and after giving some nutritious diet for three months (*a*3) and for six months (*a*6) are given below :

| | | | | | | | | |
|---|---|---|---|---|---|---|---|---|
| Before (*b*) | 40 | 60 | 55 | 65 | 43 | 70 | 80 | 60 |
| After (*a*3) | 50 | 80 | 50 | 70 | 40 | 60 | 90 | 85 |
| After (*a*6) | 70 | 90 | 100 | 90 | 75 | 65 | 70 | 120 |

Rank the values separately for each group, taking the average of the ranks for tied values.

| | | | | | | | | | | Total (R) | $R^2$ |
|---|---|---|---|---|---|---|---|---|---|---|---|
| Before(*b*) | : | 40 | 60 | 55 | 65 | 43 | 70 | 80 | 60 | | |
| Ranks | : | 1 | 1 | 2 | 1 | 2 | 3 | 2 | 1 | 13 | 169 — A |

| | | | | | | | | | | | |
|---|---|---|---|---|---|---|---|---|---|---|---|
| After (*a*3) | : | 50 | 80 | 50 | 70 | 40 | 60 | 90 | 85 | | |
| Ranks | : | 2 | 2 | 1 | 2 | 1 | 1 | 3 | 2 | 14 | 196 — B |
| After (*a*6) | : | 70 | 90 | 100 | 90 | 75 | 65 | 70 | 120 | | |
| Ranks | : | 3 | 3 | 3 | 3 | 3 | 2 | 1 | 3 | 21 | 441 — C |

Define $A = \dfrac{bk(k-1)(2k+1)}{6}$, where b is the sample size and k is the number of occasions.

Define $B = \dfrac{1}{b}\sum_{i=1}^{k} R_i^2$ .

Here the test statistic is

$$T = \frac{(b-1)\left[B - bk(k+1)^2/4\right]}{A-B}$$

Here $A = \dfrac{8\times 3(4)7}{6} = 112, B = \dfrac{1}{8}[169+169+441] = 100.75$.

$$T = \frac{7\left[100.75 - \dfrac{24\times 16}{4}\right]}{112-100.75} = 296.$$

*T* is distributed as F-distribution with (*k*-1), (*b*-1) (*k*-1) d.f. For 2, 14 d.f., at p = 0.05, critical ratio from F-Table is 3.74.

Computed *T* (=2.96) < 3.74, i.e., the differences in ranks among before, after 3 weeks and after 6 weeks are not statistically significant ($p > 0.05$).

| | |
|---|---|
| *p*-value (parametric two-way ANOVA) | : 0.006 |
| *p*-value (non-parametric two-way ANOVA) | : 0.0093 |

The difference in the p-values obtained by applying parametric and non-parametric ANOVAs may be noted. Hence, it can be seen that, parametric ANOVA should not be applied if the assumptions for applying parametric ANOVA are not satisfied.

ANOVA shows only whether the difference in the rank values among the three groups is statistically significant or not. To identify the pairs of groups (*A* and *B*, *B* and *C*, and *A* and C), which are significantly different, multiple range test should be applied. The method is explained below :

Compute $|R_b\text{-}R_{a3}|$, $|R_b\text{-}R_{a6}|$ and $|R_{a3}\text{-}R_{a6}|$.

Find $w = t_{(b\text{-}1)(k\text{-}1)} \sqrt{\left[\dfrac{2b(A-B)}{(b-1)(k-1)}\right]}$,

where $t_{(b\text{-}1)(k\text{-}1)}$ at p = 0.5 value should be obtained from Student's *t*-table (Table E) for (*b*-1) (*k*-1) d.f..

The differences indicted above are statistically significant at the chosen level of significance if they are greater than *w*.

For the example,

$|R_b - R_{a3}| = 1$, $|R_b - R_{a6}| = 8$, $|R_{a3} - R_{a6}| = 7$,

$$w = 2.145 \sqrt{\left[\frac{16 \times 11.25}{14} = 7.69\right]}.$$

All the differences are less than W. Hence the difference in ranks of IQ-values between any pair of occasions is not statistically significant.

The table below gives the *p*-values from paired *t*-test, Wilcoxon's signed rank test and multiple range tests (P and NP)

| *Group* | *T* | *WRS* | *MRP* | *MRNP* |
|---|---|---|---|---|
| Before & a3 | 0.176 | 0.1834 | > 0.05 | > 0.05 |
| Before & a6 | 0.017 | 0.0357 | < 0.05 | > 0.05 |
| a3 & a6 | 0.039 | 0.0500 | < 0.05 | > 0.05 |

The above given p = values show that the comparison of the groups applying paired *t*-test between two groups, applying Wilcoxon's signed rank test and applying parametric multiple range tests are not correct when there are more than two groups and if the assumptions for applying parametric test are not satisfied. The correct test in this situation will be multiple range test for non-parametric ANOVA.

**Table A**

| Nominal confidence level | 95% | | 99% | | Nominal confidence level | 95% | | 99% | |
|---|---|---|---|---|---|---|---|---|---|
| *n* | $R_1$ | Exact level | $R_1$ | Exact level | *n* | $R_1$ | Exact level | $R_1$ | Exact level |
| 5 | 1 | 93.8 | 1 | 93.8 | 53 | 20 | 94.7 | 18 | 98.7 |
| 6 | 1 | 96.9 | 1 | 96.9 | 54 | 20 | 96.0 | 18 | 99.1 |
| 7 | 1 | 98.4 | 1 | 98.4 | 55 | 21 | 94.2 | 18 | 99.4 |
| 8 | 2 | 93.0 | 1 | 99.2 | 56 | 21 | 95.6 | 19 | 99.0 |
| 9 | 2 | 96.1 | 1 | 99.6 | 57 | 22 | 93.7 | 19 | 99.2 |
| 10 | 2 | 97.9 | 1 | 99.8 | 58 | 22 | 95.2 | 20 | 98.8 |
| 11 | 3 | 93.5 | 2 | 98.8 | 59 | 22 | 96.4 | 20 | 99.1 |
| 12 | 3 | 96.1 | 2 | 99.4 | 60 | 23 | 94.8 | 21 | 98.7 |
| 13 | 3 | 97.8 | 2 | 99.7 | 61 | 23 | 96.0 | 21 | 99.0 |
| 14 | 4 | 94.3 | 3 | 98.7 | 62 | 24 | 94.4 | 21 | 99.3 |
| 15 | 4 | 96.5 | 3 | 98.3 | 63 | 24 | 95.7 | 22 | 98.9 |
| 16 | 5 | 92.3 | 3 | 99.6 | 64 | 25 | 94.0 | 22 | 99.2 |
| 17 | 5 | 95.1 | 4 | 98.7 | 65 | 25 | 95.4 | 23 | 98.7 |
| 18 | 5 | 96.9 | 4 | 99.2 | 66 | 26 | 93.6 | 23 | 99.1 |
| 19 | 6 | 93.6 | 4 | 99.6 | 67 | 26 | 95.0 | 23 | 99.3 |
| 20 | 6 | 95.9 | 5 | 98.8 | 68 | 26 | 96.2 | 24 | 99.0 |
| 21 | 6 | 97.3 | 5 | 99.3 | 69 | 27 | 94.7 | 24 | 99.2 |
| 22 | 7 | 94.8 | 5 | 99.6 | 70 | 27 | 95.9 | 25 | 98.8 |
| 23 | 7 | 96.5 | 6 | 98.9 | 71 | 28 | 94.3 | 25 | 99.1 |
| 24 | 8 | 93.6 | 6 | 99.3 | 72 | 28 | 95.6 | 26 | 98.7 |
| 25 | 8 | 95.7 | 7 | 98.5 | 73 | 29 | 94.0 | 26 | 99.0 |
| 26 | 8 | 97.1 | 7 | 99.1 | 74 | 29 | 95.3 | 26 | 99.3 |
| 27 | 9 | 94.8 | 7 | 99.4 | 75 | 29 | 96.3 | 27 | 98.9 |
| 28 | 9 | 96.4 | 8 | 98.7 | 76 | 30 | 95.0 | 27 | 99.2 |
| 29 | 10 | 93.9 | 8 | 99.2 | 77 | 30 | 96.0 | 28 | 98.8 |
| 30 | 10 | 95.7 | 8 | 99.5 | 78 | 31 | 94.6 | 28 | 99.1 |
| 31 | 10 | 97.1 | 9 | 98.9 | 79 | 31 | 95.8 | 29 | 98.7 |
| 32 | 11 | 95.0 | 9 | 99.3 | 80 | 32 | 94.3 | 29 | 99.0 |
| 33 | 11 | 96.5 | 10 | 98.6 | 81 | 32 | 95.5 | 29 | 99.3 |
| 34 | 12 | 94.2 | 10 | 99.1 | 82 | 33 | 94.0 | 30 | 98.9 |
| 35 | 12 | 95.9 | 10 | 99.4 | 83 | 33 | 95.2 | 30 | 99.2 |
| 36 | 13 | 93.5 | 11 | 98.9 | 84 | 33 | 96.2 | 31 | 98.8 |
| 37 | 13 | 95.3 | 11 | 99.2 | 85 | 34 | 95.0 | 31 | 99.1 |
| 38 | 13 | 96.6 | 12 | 98.6 | 86 | 34 | 96.0 | 32 | 98.7 |
| 39 | 14 | 94.7 | 12 | 99.1 | 87 | 35 | 94.7 | 32 | 99.0 |
| 40 | 14 | 96.2 | 12 | 99.4 | 88 | 35 | 95.8 | 32 | 99.3 |
| 41 | 15 | 94.0 | 13 | 98.8 | 89 | 36 | 94.4 | 33 | 98.9 |
| 42 | 15 | 95.6 | 13 | 99.2 | 90 | 36 | 95.5 | 33 | 99.2 |
| 43 | 16 | 93.4 | 14 | 98.6 | 91 | 37 | 94.1 | 34 | 98.9 |
| 44 | 16 | 95.1 | 14 | 99.0 | 92 | 37 | 95.3 | 34 | 99.1 |
| 45 | 16 | 96.4 | 14 | 99.3 | 93 | 38 | 93.9 | 35 | 98.8 |
| 46 | 17 | 94.6 | 15 | 98.9 | 94 | 38 | 95.1 | 35 | 99.0 |
| 47 | 17 | 96.0 | 15 | 99.2 | 95 | 38 | 96.0 | 35 | 99.3 |
| 48 | 18 | 94.1 | 16 | 98.7 | 96 | 39 | 94.8 | 36 | 99.0 |
| 49 | 18 | 95.6 | 16 | 99.1 | 97 | 39 | 95.8 | 36 | 99.2 |
| 50 | 19 | 93.5 | 16 | 99.3 | 98 | 40 | 94.6 | 37 | 98.9 |
| 51 | 19 | 95.1 | 17 | 98.9 | 99 | 40 | 95.6 | 37 | 99.1 |
| 52 | 19 | 96.4 | 17 | 99.2 | 100 | 41 | 94.3 | 38 | 98.8 |

**Table B1 :** WSR Test (level = 0.05)

| $n_1 \rightarrow$ / $n_2 \downarrow$ | 2 | 3 | 4 | 5 | 6 | 7 | 8 | 9 | 10 | 11 | 12 | 13 | 14 | 15 |
|---|---|---|---|---|---|---|---|---|---|---|---|---|---|---|
| 4 | | | 10 | | | | | | | | | | | |
| 5 | | 6 | 11 | 17 | | | | | | | | | | |
| 6 | | 7 | 12 | 18 | 26 | | | | | | | | | |
| 7 | | 7 | 13 | 20 | 27 | 36 | | | | | | | | |
| 8 | 3 | 8 | 14 | 21 | 29 | 38 | 49 | | | | | | | |
| 9 | 3 | 8 | 15 | 22 | 31 | 40 | 51 | 63 | | | | | | |
| 10 | 3 | 9 | 15 | 23 | 32 | 42 | 53 | 65 | 78 | | | | | |
| 11 | 4 | 9 | 16 | 24 | 34 | 44 | 55 | 68 | 81 | 96 | | | | |
| 12 | 4 | 10 | 17 | 26 | 35 | 46 | 58 | 71 | 85 | 99 | 115 | | | |
| 13 | 4 | 10 | 18 | 27 | 37 | 48 | 60 | 73 | 88 | 103 | 119 | 137 | | |
| 14 | 4 | 11 | 19 | 28 | 38 | 50 | 63 | 76 | 91 | 106 | 123 | 141 | 160 | |
| 15 | 4 | 11 | 20 | 29 | 40 | 52 | 65 | 79 | 94 | 110 | 127 | 145 | 164 | 185 |
| 16 | 4 | 12 | 21 | 31 | 42 | 54 | 67 | 82 | 97 | 114 | 131 | 150 | 169 | |
| 17 | 5 | 12 | 21 | 32 | 43 | 56 | 70 | 84 | 100 | 117 | 135 | 154 | | |
| 18 | 5 | 13 | 22 | 33 | 45 | 58 | 72 | 90 | 103 | 121 | 139 | | | |
| 19 | 5 | 13 | 23 | 34 | 46 | 60 | 74 | 93 | 107 | 124 | | | | |
| 20 | 5 | 14 | 24 | 35 | 48 | 62 | 77 | 95 | 110 | | | | | |
| 21 | 6 | 14 | 25 | 37 | 50 | 64 | 79 | | | | | | | |
| 22 | 6 | 15 | 26 | 38 | 51 | 66 | 82 | | | | | | | |
| 23 | 6 | 15 | 27 | 39 | 53 | 68 | | | | | | | | |
| 24 | 6 | 16 | 28 | 40 | 55 | | | | | | | | | |
| 25 | 6 | 16 | 28 | 42 | | | | | | | | | | |
| 26 | 7 | 17 | 29 | | | | | | | | | | | |
| 27 | 7 | 17 | | | | | | | | | | | | |
| 28 | 7 | | | | | | | | | | | | | |

**Table $B_2$ :** WSR Test (level = 0.01)

| $n_1 \rightarrow$ / $n_2 \downarrow$ | 2 | 3 | 4 | 5 | 6 | 7 | 8 | 9 | 10 | 11 | 12 | 13 | 14 | 15 |
|---|---|---|---|---|---|---|---|---|---|---|---|---|---|---|
| 5 | | | | 15 | | | | | | | | | | |
| 6 | | | 10 | 16 | 23 | | | | | | | | | |
| 7 | | | 10 | 17 | 24 | 32 | | | | | | | | |
| 8 | | | 11 | 17 | 25 | 34 | 43 | | | | | | | |
| 9 | | 6 | 11 | 18 | 26 | 35 | 45 | 56 | | | | | | |
| 10 | | 6 | 12 | 19 | 27 | 37 | 47 | 58 | 71 | | | | | |
| 11 | | 6 | 12 | 20 | 28 | 38 | 49 | 61 | 74 | 87 | | | | |
| 12 | | 7 | 13 | 21 | 30 | 40 | 51 | 63 | 76 | 90 | 106 | | | |
| 13 | | 7 | 14 | 22 | 31 | 41 | 53 | 65 | 79 | 93 | 109 | 125 | | |
| 14 | | 7 | 14 | 22 | 32 | 43 | 54 | 67 | 81 | 96 | 112 | 129 | 147 | |
| 15 | | 8 | 15 | 23 | 33 | 44 | 56 | 70 | 84 | 99 | 115 | 133 | 151 | 171 |
| 16 | | 8 | 15 | 24 | 34 | 46 | 58 | 72 | 86 | 102 | 119 | 137 | 155 | |
| 17 | | 8 | 16 | 25 | 36 | 47 | 60 | 74 | 89 | 105 | 122 | 140 | | |
| 18 | | 8 | 16 | 26 | 37 | 49 | 62 | 76 | 92 | 108 | 125 | | | |
| 19 | 3 | 9 | 17 | 27 | 38 | 50 | 64 | 78 | 94 | 111 | | | | |
| 20 | 3 | 9 | 18 | 28 | 39 | 52 | 66 | 81 | 97 | | | | | |
| 21 | 3 | 9 | 18 | 29 | 40 | 53 | 68 | 83 | | | | | | |
| 22 | 3 | 10 | 19 | 29 | 42 | 55 | 70 | | | | | | | |
| 23 | 3 | 10 | 19 | 30 | 43 | 57 | | | | | | | | |
| 24 | 3 | 10 | 20 | 31 | 44 | | | | | | | | | |
| 25 | 3 | 11 | 20 | 32 | | | | | | | | | | |
| 26 | 3 | 11 | 21 | | | | | | | | | | | |
| 27 | 4 | 11 | | | | | | | | | | | | |
| 28 | 4 | | | | | | | | | | | | | |

$n_1$ and $n_2$ are the numbers of cases in the two groups. If the groups are unequal in size, $n_1$ refers to the smaller.

**Table C :** Sum of ranks at 2 levels for WSR Test (5% and 1%)

Two-sided test

| *Number of Pairs* | *0.05* | *0.01* | *Number of Pairs* | *0.05* | *0.01* |
|---|---|---|---|---|---|
| 7 | 2 | 0 | 12 | 14 | 7 |
| 8 | 2 | 0 | 13 | 17 | 10 |
| 9 | 6 | 2 | 14 | 21 | 13 |
| 10 | 8 | 3 | 15 | 25 | 16 |
| 11 | 11 | 5 | 16 | 29 | 19 |

These numbers or smaller indicate rejection.

**Table D :** $\chi^2$-distribution

| | *α level* | | | | *α level* | | |
|---|---|---|---|---|---|---|---|
| *d.f.* | *0.10* | *0.05* | *0.01* | *d.f.* | *0.10* | *0.05* | *0.01* |
| 1 | 2.71 | 3.84 | 6.63 | 23 | 32.01 | 35.17 | 41.64 |
| 2 | 4.61 | 5.99 | 9.21 | 24 | 33.20 | 36.42 | 42.98 |
| 3 | 6.25 | 7.81 | 11.34 | 25 | 34.38 | 37.65 | 44.31 |
| 4 | 7.78 | 9.49 | 13.28 | 30 | 40.26 | 43.77 | 50.89 |
| 5 | 9.24 | 11.07 | 15.09 | 35 | 46.06 | 49.80 | 57.34 |
| 6 | 10.64 | 12.59 | 16.81 | 40 | 51.81 | 55.76 | 63.69 |
| 7 | 12.02 | 14.07 | 18.48 | 45 | 57.51 | 61.66 | 69.96 |
| 8 | 13.36 | 15.51 | 20.09 | 50 | 63.17 | 67.50 | 76.15 |
| 9 | 14.68 | 16.92 | 21.67 | 60 | 74.40 | 79.08 | 88.38 |
| 10 | 15.99 | 18.31 | 23.21 | 70 | 85.53 | 90.53 | 100.43 |
| 11 | 17.28 | 19.68 | 24.72 | 80 | 96.58 | 101.88 | 112.33 |
| 12 | 18.55 | 21.03 | 26.22 | 90 | 107.51 | 113.15 | 124.12 |
| 13 | 19.81 | 22.36 | 27.69 | 100 | 118.50 | 124.34 | 135.81 |
| 14 | 21.06 | 23.68 | 29.14 | 200 | 226.02 | 233.99 | 249.45 |
| 15 | 22.31 | 25.00 | 30.58 | 300 | 331.79 | 341.39 | 359.91 |
| 16 | 23.54 | 26.30 | 32.00 | 400 | 436.65 | 447.63 | 468.73 |
| 17 | 24.77 | 27.59 | 33.41 | 500 | 540.93 | 553.13 | 576.50 |
| 18 | 25.99 | 28.87 | 34.81 | 600 | 644.80 | 658.09 | 683.52 |
| 19 | 27.20 | 30.14 | 36.19 | 700 | 748.36 | 762.66 | 789.98 |
| 20 | 28.41 | 31.41 | 37.57 | 800 | 851.67 | 866.91 | 895.99 |
| 21 | 29.62 | 32.67 | 38.93 | 900 | 954.78 | 970.90 | 1001.63 |
| 22 | 30.81 | 33.92 | 40.29 | 1000 | 1057.72 | 1074.68 | 1106.97 |

[This table was produced by the author using SYSTAT's Data Basic]

All probabilities in this table refer to the extreme tail (right tail). Suppose you were analyzing some categorical data and that there were five degrees of freedom in the residuals. This might happen if you were looking to see if there was an association in a 2 × 6 contingency table (one variable has two possible values, the other six). If you were using the α = 0.05 level, then you would need a $\chi^2$ value and of 11.07. The area beyond this point on the graph is 5% of the total area under the curve. Only 5% of the time would you expect a value this high (or higher) if there is no association in the population (providing proper sampling and procedures are used).

**Table E** : *t*-distribution

| | Two tailed level-α | | | | | | One tailed level-α | | | | |
|---|---|---|---|---|---|---|---|---|---|---|---|
| *df* | *0.001* | *0.010* | *0.050* | *0.100* | *0.200* | *0.500* | *0.001* | *0.010* | *0.050* | *0.100* | *0.200* |
| 2 | 31.60 | 9.92 | 4.30 | 2.92 | 1.89 | 0.82 | 22.33 | 6.96 | 2.92 | 1.89 | 1.06 |
| 3 | 12.92 | 5.84 | 3.18 | 2.35 | 1.64 | 0.76 | 10.21 | 4.54 | 2.35 | 1.64 | 0.98 |
| 4 | 8.61 | 4.60 | 2.78 | 2.13 | 1.53 | 0.74 | 7.17 | 3.75 | 2.13 | 1.53 | 0.94 |
| 5 | 6.87 | 4.03 | 2.57 | 2.02 | 1.48 | 0.73 | 5.89 | 3.36 | 2.02 | 1.48 | 0.92 |
| 6 | 5.96 | 3.71 | 2.45 | 1.94 | 1.44 | 0.72 | 5.21 | 3.14 | 1.94 | 1.44 | 0.91 |
| 7 | 5.41 | 3.50 | 2.36 | 1.89 | 1.41 | 0.71 | 4.79 | 3.00 | 1.89 | 1.41 | 0.90 |
| 8 | 5.04 | 3.36 | 2.31 | 1.86 | 1.40 | 0.71 | 4.50 | 2.90 | 1.86 | 1.40 | 0.89 |
| 9 | 4.78 | 3.25 | 2.26 | 1.83 | 1.38 | 0.70 | 4.30 | 2.82 | 1.83 | 1.38 | 0.88 |
| 10 | 4.59 | 3.17 | 2.23 | 1.81 | 1.37 | 0.70 | 4.14 | 2.76 | 1.81 | 1.37 | 0.88 |
| 11 | 4.44 | 3.11 | 2.20 | 1.80 | 1.36 | 0.70 | 4.02 | 2.72 | 1.80 | 1.36 | 0.88 |
| 12 | 4.32 | 3.05 | 2.18 | 1.78 | 1.35 | 0.70 | 3.93 | 2.68 | 1.78 | 1.36 | 0.87 |
| 13 | 4.22 | 3.01 | 2.16 | 1.77 | 1.35 | 0.69 | 3.85 | 2.65 | 1.77 | 1.35 | 0.87 |
| 14 | 4.44 | 2.98 | 2.14 | 1.76 | 1.34 | 0.69 | 3.79 | 2.62 | 1.76 | 1.35 | 0.87 |
| 15 | 4.07 | 2.95 | 2.13 | 1.75 | 1.34 | 0.69 | 3.73 | 2.60 | 1.75 | 1.34 | 0.87 |
| 16 | 4.01 | 2.92 | 2.12 | 1.75 | 1.33 | 0.69 | 3.69 | 2.58 | 1.75 | 1.34 | 0.86 |
| 17 | 3.97 | 2.90 | 2.11 | 1.74 | 1.33 | 0.69 | 3.65 | 2.57 | 1.74 | 1.33 | 0.86 |
| 18 | 3.92 | 2.88 | 2.10 | 1.73 | 1.33 | 0.69 | 3.61 | 2.55 | 1.73 | 1.33 | 0.86 |
| 19 | 3.88 | 2.80 | 2.09 | 1.73 | 1.33 | 0.69 | 3.58 | 2.54 | 1.73 | 1.33 | 0.86 |
| 20 | 3.85 | 2.85 | 2.09 | 1.72 | 1.33 | 0.69 | 3.55 | 2.53 | 1.72 | 1.33 | 0.86 |
| 21 | 3.82 | 2.83 | 2.08 | 1.72 | 1.32 | 0.69 | 3.53 | 2.52 | 1.72 | 1.32 | 0.86 |
| 22 | 3.79 | 2.82 | 2.07 | 1.72 | 1.32 | 0.69 | 3.50 | 2.51 | 1.72 | 1.32 | 0.86 |
| 23 | 3.77 | 2.81 | 2.07 | 1.71 | 1.32 | 0.69 | 3.48 | 2.50 | 1.71 | 1.32 | 0.86 |
| 24 | 3.75 | 2.80 | 2.06 | 1.71 | 1.32 | 0.68 | 3.47 | 2.49 | 1.71 | 1.32 | 0.86 |
| 25 | 3.73 | 2.79 | 2.06 | 1.71 | 1.32 | 0.68 | 3.45 | 2.49 | 1.71 | 1.32 | 0.86 |
| 30 | 3.65 | 2.75 | 2.04 | 1.70 | 1.31 | 0.68 | 3.39 | 2.46 | 1.70 | 1.31 | 0.85 |
| 35 | 3.59 | 2.72 | 2.03 | 1.69 | 1.31 | 0.68 | 3.34 | 2.44 | 1.69 | 1.31 | 0.85 |
| 40 | 3.55 | 2.70 | 2.02 | 1.68 | 1.30 | 0.68 | 3.31 | 2.42 | 1.68 | 1.30 | 0.85 |
| 45 | 3.52 | 2.69 | 2.01 | 1.68 | 1.30 | 0.68 | 3.28 | 2.41 | 1.68 | 1.30 | 0.85 |
| 50 | 3.50 | 2.68 | 2.01 | 1.68 | 1.30 | 0.68 | 3.26 | 2.40 | 1.68 | 1.30 | 0.85 |
| 60 | 3.46 | 2.66 | 2.00 | 1.67 | 1.30 | 0.68 | 3.23 | 2.39 | 1.67 | 1.30 | 0.85 |
| 70 | 3.44 | 2.65 | 1.99 | 1.67 | 1.29 | 0.68 | 3.21 | 2.38 | 1.67 | 1.29 | 0.85 |
| 80 | 3.42 | 2.64 | 1.99 | 1.66 | 1.29 | 0.68 | 3.20 | 2.37 | 1.66 | 1.29 | 0.85 |
| 90 | 3.40 | 2.63 | 1.99 | 1.66 | 1.29 | 0.68 | 3.18 | 2.37 | 1.66 | 1.29 | 0.85 |
| 100 | 3.39 | 2.63 | 1.98 | 1.66 | 1.29 | 0.68 | 3.17 | 2.36 | 1.66 | 1.29 | 0.85 |
| ∞ | 3.29 | 2.58 | 1.96 | 1.64 | 1.15 | 0.67 | 3.09 | 2.33 | 1.64 | 1.28 | 0.67 |

[This table was produced by the author using SYSTAT's Data Basic]

## References

Armitage, P. and Berry, G. (1978), *Statistical Methods in Medical Research,* Blackwell Scientific Publications, London.

Bourke, G.J., Daly, I.E. and Mcgilvray, J. (1988), *Interpretation and Uses of Medical Statistics,* Blackwell Scientific Publications, London.

Castle, W.M. (1984), *Statistics in Small Doses,* Churchill, Livingstone, Edinburgh.

Colton, T. (1974), *Statistics in Medicine,* Little Brown & Co., Boston.

Colquhoun, D. (1971), *Lectures on Biostatistics – An Introduction to Statistics with Applications in Biology and Medicine,* Clarendon Press Oxford.

Conover, W.J. (1980), *Practical Non-parametric Statistics,* 2nd Ed., John Wiley, NY.

Feinstein, A.R. (1977), *Clinical Biostatistics,* The C.V. Mosby & Co.

Matthews, D.E. and Farewell, V.(1985), *Using and Undertaking Medical Statistics,* Karger, London.

Petrie, A. (1978), *Lecture Notes on Medical Statistics,* Blackwell Scientific Publications, Oxford.

Pett, M.A. (1997), *Non-parametric Statistics for Health Care Research,* Sage Publications, London.

Schor, S.S. (1968), *Fundamentals of Biostatistics,* G.P. Putnam & Sons, New York.

Siegel, S. (1956), *Non-parametric Statisitcs for Behavioural Sciences,* McGraw-Hill, NY.

Sokal, R. R. and Rohlf, F.J. (1987), *Introduction to Biostatistics,* W.H. Freeman, N.Y.

# 6

# Use of Hierarchical Models in Analysis of Public Health Data

*Sada Nand Dwivedi*

## 1. Introduction

A number of micro as well as macro level analyses are reported on various public health issues as an attempt to understand its regional epidemiology. But, most of the analyses dealing with identification of associated factors do not make efforts to consider hierarchical structure of the data. They simply disaggregate higher-level variables at lower level and carry out analysis through traditional methods that need assumption of independence of records. However, disaggregation under traditional methods of data analysis violates this assumption, and that results into underestimation of standard error of estimates providing significance of some of irrelevant covariates. Similarly, aggregation of lower level variables at higher level also provides distorted results under traditional methods. To overcome this problem, models using multilevel analysis may be appropriate that takes hierarchical structure into account. This procedure makes it possible to incorporate variables from all levels at their own level to get correct analysis and proper interpretation of the data on current contraceptive use. Unobserved community effects are also taken into account. It also provides the relative importance of characteristics related to various levels, which provides important clues strengthening ensuing public health programmes. The present paper will discuss about the multilevel analysis. For this, for example, the data of a most populous Indian State "Uttar Pradesh (UP)", covered under National Family Health Survey (NFHS) conducted during 1998-1999, will be used. The sample design adopted for the NFHS is a systematic, two-stage-stratified sample of households. The NFHS in UP is a state representative survey of ever-married women age 15-49. The main objective of the survey was to collect reliable and up-to-date information on family planning, fertility, mortality, and maternal and child health providing state-level estimates. Its important objective was to provide high quality data to academicians and researchers for undertaking analytical research. For analysis, currently married women who were neither pregnant nor continuing with post-partum amenorrhoea (PPA) will be considered. A two-level logistic regression analysis may be carried out for which women's level (level 1) and PSU (Primary Sampling Unit) level (level 2) variables may be considered. Women whose records were complete in relation to all variables considered in analysis may only be included in the analysis. Women with incomplete records may however be negligible. In view of poor contraceptive adoption and majority of couples going for sterilization in Uttar Pradesh and also taking into account the results from a series of exploratory models, to have meaningful observations, the current contraceptive use (including sterilization, modern methods etc.) may be considered as dependent variable.

## 2. The Model

Under many areas of public health research including contraceptive adoption, the data structures are often hierarchical in nature. Till recently, there have been generally two classical statistical procedures to deal with them. The first is to disaggregate all higher order variables to the individual level and carry out the analysis at individual level. Thus, the assumption of independence of observations that is basic for the classical statistical techniques becomes invalid. On the other hand, the second is to aggregate the individual level variables to the higher level and do the analysis at higher level. Thus, all the within group information are thrown away, which may be as much as 80% or 90% of the total variation, before the analysis is carried out. As a consequence, relations between aggregated variables are often much stronger giving distorted interpretation at individual level (Bryk and Roudenbush, 1992, Goldstein, 1995, Drevett and Goldstein, 1993, Goldstein et al., 1994, Fiona and Ian, 1996, Curtis et al., 1993, Goldstein and Healy, 1995, Fiona et al., 1996). Also, a number of quality/ behavioural characteristics may not be either available or adequately measured, particularly the quality / behaviour as perceived by the client/ community. Almost all survey data sets involve some type of cluster sampling where groups of individuals from randomly chosen segments of a population are interviewed. But, generally, statistical methods are used which assume that the data was obtained from a simple random sample. A number of recent papers in the epidemiologic literature contend that ignoring above described issues (unobserved community effects) in the analysis of data produces downward bias in the standard errors of the estimated parameters, which leads to erroneous estimates of the impact of individual variables (Drevett and Goldstein, 1993, Goldstein et al., 1994, Fiona and Ian, 1996). However, these issues, which are considered to be of great importance in epidemiological studies, are not addressed in analysis quite often.

This study deals with application of multilevel analysis of data on contraceptive adoption in "Uttar Pradesh" that takes hierarchical structure into account and that makes it possible to incorporate variables from all levels which leads to correct analysis and proper interpretation of the data. It also provides the relative importance of a woman's individual characteristics and those of the area in which she lives, which provides important clues strengthening ensuing public health programmes.

### 2.1 The Data

The data used in the present study are from the National Family Health Survey (NFHS-II) conducted in Uttar Pradesh. The sample design adopted for the NFHS-II is also a systematic, multi-stage-stratified sample of households. The NFHS-II in UP is a state representative survey of ever-married women aged 13-49 years. The main objective of the survey was to collect reliable and up-to-date information on family planning, fertility, mortality, and maternal and child health providing state-level estimates. Its important objective was to provide high quality data to academicians and researchers for undertaking analytical research (NFHS Report, 2000, UP-NFHS Report, 2001). The NFHS in Uttar Pradesh was conducted on 266 rural areas Primary Sampling Units (PSUs) and 67 urban area CEBs. In rural areas, the 1991 Census list of villages served as the sampling frame, and a two-stage sample design was adopted with the selection of villages (PSUs) in the first stage, households in selected villages in the next stage. On the other hand, in urban areas, the list of urban wards/Census Enumeration Blocks provided by the Registrar General of India for 1991 served as the sampling frame. Accordingly, a three-stage sample design was adopted : selection of wards (PSU), Census Enumeration Blocks (CEB) in selected wards followed by selection of households in each of the selected CEBs. The selection of PSUs/CEBs was systematic with probability proportional to size (PPS). The households to be interviewed were selected from the household lists using systematic sampling with equal probability. On an average, in rural area, 30 households were selected from each selected PSU. In urban area also, on an average, 30 households per CEB were selected. A total of 9292 ever-married women of age 15-49 years from 8682 households were interviewed. More details are available in the state level report for Uttar Pradesh released in April, 2001(UP-NFHS Report, 2001).

For the present study, all the currently married women who were neither pregnant nor continuing with post-partum amenorrhoea (ppa) were considered. Women whose records were complete in relation to all variables considered in analysis could only be included in the analysis. Women with incomplete records were however negligible (less than one per thousand). Accordingly, 5028 currently married women were included in analyses that were neither pregnant nor continuing with post-partum amenorrhoea (PPA). In view of poor contraceptive adoption and majority of couples going for sterilization in Uttar Pradesh and also taking into account the results from a series of exploratory models, to have meaningful observations, the current contraceptive use (including sterilization, modern methods etc.) was considered to be the dependent variable. Similarly, taking into account theoretical consideration as well as the results from a series of exploratory models, woman's level variables considered in the analysis were present age in years; marital duration in years; residential status (Rural/Urban); religion (Hindu/Others); type of house (Kuchcha + Semi-Pucca/ Pucca); electricity in the house (No/Yes); hearing of TV message (No/Yes); working status (No/Yes); education (Illiterate/Primary School/ Middle School/High School and above); husband's education (Illiterate/1-4 years/ 5-7 years/ 8-10 years/11-12 years/13 years and above); number of living sons (0/ 1/ 2/ 3 and above) and number of ever born male children (0/ 1/ 2/ 3/ 4 and above). The survey, in addition to the standard set of questions on individual level socio-demographic and other variables, included questions on higher-level, viz., households/ PSU/ district/ state. PSU level data could also be organized and considered in the analysis. These were availability of all weather roads in the locality (No/ Yes); distance of locality from primary school (2 kms. and above/ less than 2 kms.) and distance of locality from health centre (3 kms. and above/ less than 3 kms.).

### 2.2 The Methodology

As a special case of the general framework (Goldstein, 1995, Curtis et al., 1993), as used by others (Sajeda et al., 1996, Fiona and Ian, 1996; Dwivedi and Sundaram, 2000), this study also used multilevel logistic regression to estimate the individual and areal factors that influence current contraceptive use. The model fitted takes the form

$$\log [p_{ij} \ /(1-p_{ij})] = x_{ij}\, a + w_j\, b + u_j \ + e_{ij},$$

where $p_{ij}$ is the probability that woman i in PSU j uses some method of contraception; $x_{ij}$ and $w_j$ are vectors of individual and PSU-level characteristics respectively; and a and b are vectors of estimated parameter coefficients. $u_j$ $(\sim N(0, \sigma_u^2))$ is error term at the PSU level and $e_{ij}$ $(\sim N(0,\sigma^2))$ is error term at individual level. This model is an example of what is called a two-level model—individual woman (level 1) are nested within PSU (level 2). The purpose of this approach is to control for the correlation between women in a particular PSU. The PSU error term ($u_j$) in the model gives an indication of the variation after controlling for the individual level characteristics. In this study, the model is estimated using the computer programme MLwin for multilevel analysis (Rasbash and Woodhouse,1995). However, univariate analysis including rate ratio (RR) and 95% confidence interval (95% CI) was completed through STATA software (Stata Corporation, 1995). The results were considered significant at 5% level of significance.

## 3. The Results

All the currently married women who were neither pregnant nor continuing with post-partum amenorrhoea (ppa) were analyzed in the present study. Women whose records were complete in relation to all variables considered in analysis could only be included in the analysis. Women with incomplete records were however negligible. Thus, as per inclusion criteria, 5028 currently married women were analyzed who were neither pregnant nor continuing with post-partum amenorrhoea (PPA).

Table 6.1 demonstrates percentage of contraceptive users in every sub- categories of the considered variables. This table also displays the unadjusted risk ratio (RR) and its 95% confidence interval (95% CI) estimate of contraceptive use in various categories of a variable in comparison to reference category. The results shown in Table 6.1 clearly indicate that urban dwellers were more likely to use contraceptives than their rural counterparts (RR : 2.14; 95% CI : 1.92-2.38). Similarly, women whose houses have electricity than those staying in houses without electricity (1.59; 1.49-1.69), women getting TV message than those without message (1.65; 1.54-1.78), and working women than non-working (1.51; 1.31-1.73) were more likely to use contraceptive methods. Also, contraceptive adoption was highly associated positively with number of living sons/ number of ever born male children. In addition to wife's education, husband's education also showed a significant positive association with contraceptive adoption. But, other than Hindu women had 21% less chance (significant) for contraceptive use (Table 6.1). PSU level characteristics such as presence of all weather road (1.31; 1.23-1.40); primary school within two kilometres (1.03; 1.01-1.04); and health center within three kilometres (1.31; 1.23-1.39) showed significant positive relationships with contraceptive adoption.

**Table 6.1 :** Distribution of Currently Married Women excluding Pregnant and Continuing PPA who Practice Contraception by Selected Individual Characteristics and Related Risk Ratio (RR) and 95% Confidence Interval (95% CI)

| *Variable* | | *Users Number (%)* | *Non-Users Number (%)* | *RR (95% CI)* |
|---|---|---|---|---|
| Residence | Rural | 1543(40.0) | 2358(60.0) | 1.00 |
| | Urban | 716(63.5) | 411(36.5) | 2.14(1.92-2.38) |
| Religion | Hindu | 1981(45.9) | 2338(54.1) | 1.00 |
| | Others | 278(39.2) | 431(60.8) | 0.79(0.69-0.91) |
| Type of House | Kuchcha+ Semi pucca | 2185(61.8) | 1349(38.2) | 1.00 |
| | Pucca | 910(60.9) | 584(39.1) | 0.97(0.89-1.06) |
| Electricity | No | 1049(36.4) | 1835(63.6) | 1.00 |
| | Yes | 1210(56.4) | 934(43.6) | 1.59(1.49-1.69) |
| TV message | No | 1193(37.6) | 1978(62.4) | 1.00 |
| | Yes | 1066(57.4) | 791(42.6) | 1.65(1.54-1.78) |
| Working Status | No | 1710(42.8) | 2283(57.2) | 1.00 |
| | Yes | 549(53.1) | 486(46.9) | 1.51(1.31-1.73) |
| Education | Illiterate | 1315(40.9) | 1898(59.1) | 1.00 |
| | Primary School | 313(48.5) | 333(51.5) | 1.29(1.12-1.48) |
| | Middle School | 199(46.5) | 229(56.5) | 1.22(1.02-1.46) |
| | High School & Above | 432(58.3) | 309(41.7) | 1.77(1.55-2.02) |
| Husband's Education | Iliterate | 438(37.6) | 727(62.4) | 1.00 |
| | 1-4 years | 388(44.1) | 93(55.9) | 4.14(3.38-5.08) |
| | 5-7 years | 356(40.8) | 17(59.2) | 19.62(12.19-31.59) |
| | 8 years & above | 1077(51.1) | 1032(48.9) | 1.21(1.15-1.27) |
| Number of living sons | 0 | 125(10.2) | 1100(89.8) | 1.00 |
| | 1 | 571(41.0) | 822(59.0) | 1.92(1.80-2.04) |
| | 2 | 941(66.7) | 470(33.3) | 2.95(2.73-3.14) |
| | 3 & Above | 622(62.3) | 377(37.7) | 3.26(2.97-3.58) |
| Number of Everborn Male Children | 0 | 117(10.4) | 1005(89.6) | 1.00 |
| | 1 | 492(38.6) | 781(61.4) | 1.85(1.73-1.97) |
| | 2 | 833(63.7) | 474(36.3) | 2.74(2.53-2.96) |
| | 3 | 430(62.7) | 256(37.3) | 3.87(3.44-4.38) |
| | 4 & Above | 387(60.5) | 253(39.5) | 3.82(3.39-4.31) |
| All Weather Road in Locality | No | 1155(40.0) | 1736(60.0) | 1.00 |
| | Yes | 1104(51.7) | 1033(48.3) | 1.31(1.23-1.40) |
| Distance of Locality from Primary School | 2 & Above(kms) | 127(36.0) | 226(64.0) | 1.00 |
| | < 2 (kms) | 2132(45.6) | 2543(54.4) | 1.03(1.01-1.04) |
| Distance of Locality from Health Centre | 3 & Above (kms) | 1084 (39.4) | 1668 (60.6) | 1.00 |
| | < 3 (kms) | 1175(51.6) | 1101 (48.4) | 1.31(1.23-1.39) |
| Total | | 2259 (44.9) | 2769 (55.1) | — |

Table 6.2 embodies the two level model for contraceptive adoption dealing with women who were neither pregnant nor continuing with PPA. In other words, this table also displays the adjusted risk-ratio (RR) and its 95% confidence interval (95% CI) estimate of contraceptive use in various

categories of a variable in comparison to reference category, which were adjusted for age, marital duration and also for the remaining considered variables. The results shown in Table 6.2 clearly reveal that adjusted chance of contraceptive adoption decreased in relation to every variable other than number of living sons. This adjustment demonstrated insignificant association of contraceptive adoption in relation to husband's education till 7 years. Also, among the PSU level variables, only availability of all weather road in locality was found to contribute significantly towards contraceptive adoption even after adjustment. The PSU-level variance reported in Table 6.2, which is meant as the unexplained PSU-level variation after controlling for the considered characteristics, was found insignificant. However, this variation may be the result of variables that have not been considered, or in some cases, cannot be observed.

## 4. Conclusion

There is no need to point out all studies (through individual/district/state level analysis) dealing with differentials/determinants of contraceptive adoption, because they may run into hundreds even those reported in India. It is well-known fact that studies involving large number of important variables in the form of their proper scales of measurements combined with appropriate analytical procedure will provide more valid and stable results. The results of this study however confirm earlier findings from India that the use of contraceptives is affected by a host of individual and community characteristics.

This study provides central finding that those, who have even one living son, were seven fold most likely to adopt contraception. However, looking at the 95% C.I., the very high upper confidence limit indicates instability in adjusted risk ratio. This reaches twenty fold in case of two or more living sons. This may be possible in North Indian population

**Table 6.2:** Rate Ratio (RR) and 95% Confidence Interval (95% CI) Estimates for Two Level Logistic Model of Contraceptive Use versus Non-use using Currently Married Women but excluding Pregnant and Continuing PPA

| *Variable* | | *RR* | *95% CI* | |
|---|---|---|---|---|
| | | | *Lower* | *Upper* |
| Age (years) | | 1.02 | 0.99 | 1.05 |
| Marital Duration (yrs) | | 1.08 | 1.05 | 1.12 |
| Residence | Rural | 1.00 | — | — |
| | Urban | 1.61 | 1.25 | 2.06 |
| Religion | Hindu | 1.00 | — | — |
| | Others | 0.69 | 0.55 | 0.87 |
| Type of House | Kuchcha + Semi Pucca | 1.0 | — | — |
| | Pucca | 1.11 | 0.91 | 1.37 |
| Electricity | No | 1.0 | — | — |
| | Yes | 1.24 | 1.03 | 1.51 |
| TV Message | No | 1.0 | — | — |
| | Yes | 1.56 | 1.29 | 1.89 |
| Working Status | No | 1.0 | — | — |
| | Yes | 1.25 | 1.05 | 1.48 |
| Education | Illiterate | 1.0 | — | — |
| | Primary | 1.40 | 1.12 | 1.76 |
| | Middle | 1.67 | 1.25 | 2.23 |
| | High School & Above | 2.13 | 1.54 | 2.85 |
| Husband's Education | Illiterate | 1.0 | — | — |
| | 1-4 years | 1.38 | 0.98 | 1.95 |
| | 5-7 years | 1.21 | 0.95 | 1.54 |
| | 8-10 years | 1.35 | 1.07 | 1.71 |
| | 11-12 years | 1.38 | 1.08 | 1.76 |
| | 13 years & above | 1.65 | 1.29 | 2.12 |
| No. of living sons | 0 | 1.0 | — | — |
| | 1 | 7.05 | 3.21 | 15.5 |
| | 2 | 20.34 | 8.88 | 46.6 |
| | 3 & Above | 20.86 | 8.77 | 49.61 |
| No. of Ever born Male children | 0 | 1.0 | — | — |
| | 1 | 0.46 | 0.21 | 1.04 |
| | 2 | 0.43 | 0.18 | 1.01 |
| | 3 | 0.32 | 0.13 | 0.78 |
| | 4 & Above | 0.21 | 0.08 | 0.53 |
| All Weather Road in Locality | No | 1.0 | — | — |
| | Yes | 1.26 | 1.11 | 1.43 |
| Distance of Locality from Primary School | 2 & More (kms) | 1.0 | — | — |
| | < 2 (kms) | 0.50 | 0.18 | 1.35 |
| Distance of Locality from Health centre | 3 & more (kms) | 1.0 | — | — |
| | < 3 (kms) | 3.61 | 0.40 | 26.20 |
| Random effects | | | | |
| PSU Level Variance | | 0.2294 (.1264) | | |

because of their unique thinking of insecurity. Also, those were more likely to adopt contraception who were educated preferably high school and above; whose husbands were literate preferably schooling of 8 years and above; and who were exposed to TV message. Muslim and other religious women were less likely than Hindu to adopt contraception.

The PSU-level availability of all weather road was positively associated with contraceptive adoption. This may be attributed to the fact that availability of all weather road may be an indication towards better supply as well as the use of facilities related to family welfare programme and health. The PSU level variance, which is the unexplained PSU-level variation after controlling for the considered characteristics, was found insignificant. There is, however, need to consider some more PSU-level variables. Also, consideration of district level variables in the analysis may further refine the results. These analyses do suggest that contraceptive adoption depends on the current levels of willingness to accept and to institute change in a community, which may be affected by literacy and utilization of health facilities.

## Acknowledgements

The authors are thankful to the All India Institute of Medical Sciences, New Delhi, for its timely financial support and other facilities, which were necessary to carry out this study. Finally, Dr. Rajvir Singh, Department of Biostatistics, AIIMS, New Delhi and Dr. D. Sahu, Institute for Research in Medical Statistics, New Delhi also deserve thanks for their help in data analysis.

## References

Bryk, A.S., Raudenbush, S.W. (1992), *Hierarchical Linear Models: Applications and Data Analysis Methods*, Sage Publications.

Curtis, S.L., Ian, D. and McDonald J. W. (1993) Birth Interval and Family Effects on Post Neonatal Mortality in Brazil, *Demography*, 30(1), 33-43.

Drevett, R.F. and Goldstein, H. (1993), Modeling Lactation using an Inverse Polynomial in a Multilevel Statistical Model, *Statistics in Medicine*, 12, 949-954.

Dwivedi, S.N. (1992), Contribution of Some Socio-economic

Variables towards Explaining the Level of Adoption of Various Family Planning Devices in India during 1987, *Demography India*, 21(2), 239-245.

Dwivedi, S.N. and Sundaram, K.R. (2000), Epidemiological Models and Related Simulation Results for Understanding of Contraceptive Adoption in India, *Intl. J. Epidemiology*, 29, 300-307.

Evaluation Project. UP district level baseline surveys. The Evaluation Project, Carolina Population Center, Chapel Hill, USA, 1996.

Fiona, S. and Ian, D. (1996), Immunization Uptake in Rural Bangladesh : A Multilevel Analysis, *J. Roy. Statist. Soc., Series* B, 57, 289-299.

Fiona, S., Ian, D., and Wang, D. (1996), The Determinants of Contraceptive Use in China: A Multilevel Multinomial Discrete Hazards Modelling Approach, *Demography*, 33(1),12-24.

Goldstein, H. (1995), *Multilevel Statistical Models*, London: Edward Arnold.

Goldstein, H. and Healy, M.J.R. (1995), The Graphical Presentation of a Collection of Means, *Journal of the Royal Statistical Society*, Series A, 158, 175-177.

Goldstein, H., Healy M.J.R. and Rasbash, J.(1994), Multilevel Time Series Models with Applications to Repeated Measures Data, *Statistics in Medicine*,13,1643-1655.

Health Information of India. An Annual Publication of the Directorate General of Health Services, Government of India 1999, ISSN 0971-0159.

IFPS Project. Innovations in family planning services project. USAID/India Program Summary, USAID, New Delhi, 1995.

MRHS Survey. Uttar Pradesh : Male Reproductive Health Survey 1995-1996. The Evaluation Project, Carolina Population Center, Chapel Hill, USA, 1996.

NFHS : A Final Report of the "National Family Health Survey, 1998-99". International Institute for Population Sciences, Bombay (Mumbai, at present), India, 2000.

Pathak, K.B. and Prasad, C.V.S. (1977), Component Analysis of the Correlations of Some Factors with Family Planning Performance by Methods in India during 1967-73, *Journal of Family Welfare*, XXIII(3), 3-14.

PERFORM Survey. A Final Report on the Performance Indicators for the Innovations in Family Planning Services Project 1995.

The Evaluation Project, Carolina Population Center, Chapel Hill, USA,1996.

Pritchett, L.H. (1994), Desired Fertility and the Impact of Population Policies, *Population and Development Review*, 20(1), 1-55.

Rasbash, J. and Woodhouse, G. (1995), MLwin Command Reference. London : Institute of Education.

Robey, B., Rutstein, S.O. and Morris, L. (1993), The Fertility Decline in Developing Countries, *Scientific American*, 269(6), 60-66.

Sajeda, A., Ian, D. and Fiona, S. (1996), Contraception and Religious Practice in Bangladesh, The Population Council, New York, NY, *Working Paper* No. 83, 1-35.

Scribner, S. (1994), Policies Affecting Fertility and Contraceptive Use : An Assessment of Twelve Sub-Saharan Countries, World Bank Discussion Paper No.259, Africa Technical Department Series. World Bank, Washington, DC, 1994.

Stata Corp. Stata Statistical Software. Release 4.0. College Station, Texas : Stata Corporation, 1995.

UP-NFHS : Uttar Pradesh : National Family Health Survey, 1998-99 : A Final Report. International Institute for Population Sciences, Bombay (Mumbai, at present), India, 2001.

7

# Joint Distribution of Ages at Marriage and Ages at Sterilization

*M. Vivekananda Murty*

## 1. Introduction

Many demographic studies established the fact that age at marriage has a significant impact on fertility behaviour. In Asian countries, over recent years, a transition has been taken place from traditional style of living to modern style of living. This transition has a greater impact on the patterns of marriage. A major curser of the process is the trend towards late marriage. The age at marriage in a country like India has a tremendous impact on fertility because procreation occurs totally after marriage. Thus in India one of the important determinants of fertility control is marriage. Hence, age at marriage is an important demographic variable to be studied.

Glick (1967) observed that age at marriage and spacing of children are among the variables, which help to explain human fertility patterns. Nevett (1967) observed early marriage in India is a factor towards large family. Husain (1971) studied the relationship between general fertility and mean age of marriage. Anderson et al. (1987) developed a proportionalized hazard model for timing of age at marriage of women in Malaysia. Nair and Koteshwar (1987) studied the trends in female age at marriage in rural Northern Karnataka. Ramabhadran (1987) examined the effect of age at marriage of women in India on fertility and mortality rates. Sinha (1987) established an inversion relationship between women age at marriage and their fertility. Goyal (1989) studied mean marriage age of female in India and suggested that research is required on identifying the correlates, which determine the marriage age in order to develop appropriate policy and programmes. Karkal and Rajan (1989) reviewed recent trends in age at marriage in India and observe that the age at marriage of girls has been showing an increasing trend over the years. Nath et al. (1992) suggested that female age at marriage is an important variable in the human reproduction process especially in traditional societies.

Audinarayana (1993) studied the influence of some variables on age at marriage and also on the gap between menarche and marriage. The age at marriage of a woman in India is traditionally low and has shown a very marginal increase over the last five decades. This shift is comparatively lesser than that of our Asian neighbouring countries like, Indonesia, Sri Lanka and Korea. Several researchers conducted empirical studies regarding relation between the fertility and age at marriage.

Another factor that influences the fertility is the practice of family planning methods. The termination method known as sterilization has a significant impact on population growth. Generally, the terminal methods will be accepted by one of the couple after they reached a satisfied level of the family size. In India ages at marriage are varying over different regions. Religions and cultural variations are also significant as far as age at marriage is concerned. So the adoption of Family planning methods has wide variation across regions and also among different population groups. In regions where fertility decline is fast with unfavourable social and economic conditions like, low female literacy rate and low standard of living, the gap between age at marriage and age at sterilization is getting narrowed down. This situation is prompted, because of rigorous Family planning programmes pursued by the various governments. The state of Andhra Pradesh in India is a classic example of this situation.

Another important fact is that the age at marriage and age at sterilization are also very much influenced by different social, economical and educational levels, which are random in nature. Hence these two variables : age at marriage and age at sterilization are random in nature. This leads to statistical treatment of study for these variables and their inter-relationship.

To study the behaviour and trends of these variables (age at marriage and age at sterilization), it is necessary to develop suitable probability models (distributions) governed by these variables. These models were analyzed with the help of the data drawn during 1998-1999 in Rapid Household Survey (RHS) under Reproductive and Child Health (RCH) survey for Andhra Pradesh.

## 2. Probability Distribution of Age at Marriage

Marriage is one of the important events in any individual's life irrespective of religion she/he belongs to. All religions attach great importance to marriage. That is, marriage is an important religious duty as well as an important social event in one's life. Marriage is influenced by a host of cultural, social and economic factors. The age at marriage of the woman is often determined by interplay of these multiple factors. Cultural obligations, social customs and economic conditions at macro level influence the age at marriage of the woman. An individual or micro-level studies on reasons for early marriage point out that the fear of loss of virginity of the girl before marriage in the rural settings greatly influence the parents to perform the marriage of the girl immediately after she attains puberty. Illiteracy and rise in cost of the marriage including dowry are some of the other factors influencing the age at marriage. In traditional societies the effects of modernization percolates slowly particularly in rural areas and hence the slower improvement in age at marriage in the rural areas. However, in the urban settings the effects of education and influence of other mass media play a dominant role in getting the marriages at higher ages. Hence, while attributing a statistical distribution to the age at marriage one has to postulate a mixture model. To fit a probability model to the variable under consideration one has to think about a continuous positive range distribution having long upper tail and positively skewed. In both the situations, the age at marriage is to be approximated with a left truncated distribution.

The development of this theory was made analogues to reliability analysis by treating the age at marriage as equivalent to that of failure time in reliability analysis as marriage can be considered a failure of celibacy and open for sexual life. The Weibull distribution is very flexible and extensively used in life testing as a failure time distribution. It includes constant rate, increasing rate and decreasing rate of change. Therefore, it is postulated that a mixture of left truncated Weibull distribution is better suited for the variable under study.

A sample of 5,212 married women in Andhra Pradesh is collected and presented in Table 7.1.

**Table 7.1:** Bivariate data on Age at Marriage and Age at Sterilization in Andhra Pradesh

| *X* \ *Y* | *16 - 19* | *19 - 22* | *22 - 25* | *25 - 28* | *28 - 31* | *31 - 34* | *34 - 37* | *Total* |
|---|---|---|---|---|---|---|---|---|
| 13 – 16 | 139 | 988 | 759 | 601 | 234 | 5 | - | 2,726 |
| 16 – 19 | - | 157 | 951 | 657 | 141 | 6 | - | 1,912 |
| 19 – 22 | - | - | 225 | 139 | 49 | 27 | - | 440 |
| 22 – 25 | - | - | - | 58 | 29 | 13 | - | 100 |
| 25 – 28 | - | - | - | - | 11 | 15 | 1 | 27 |
| 28 – 31 | - | - | - | - | - | 3 | 3 | 6 |
| 31 – 34 | - | - | - | - | - | - | 1 | 1 |
| Total | 139 | 1,145 | 1,935 | 1,455 | 464 | 69 | 5 | 5,212 |

From the frequency curve of the data, it observed that the age at marriage is a bimodal having one mode at 15 years and another at 18 years, and resembles like a mixture of left truncated Weibull distribution. This leads to the use of a two-component mixture model for the age at marriage, which is having two peaks, one may be due to the population in rural culture and the other due to the population having urban culture.

The probability density function of the left truncated mixture of Weibull distribution is given by

$$f(x) = p\frac{c_1}{a_1}\left(\frac{x-b}{a_1}\right)^{c_1-1} e^{-\left(\frac{x-b}{a_1}\right)^{c_1}} +$$

$$q\frac{c_2}{a_2}\left(\frac{x-b}{a_2}\right)^{c_2-1} e^{-\left(\frac{x-b}{a_2}\right)^{c_2}}, \; x > b \qquad \ldots (2.1)$$

$$0 < p < 1; \; q = 1 - p; \; a_1, c_1, a_2, c_2 > 0.$$

The maximum likelihood estimates of the parameters are $p = 0.393, q = 0.607, a_1 = 6.3, c_1 = 5.3, a_2 = 2.3, c_2 = 3.98$ and $b = 13$ for Andhra Pradesh state as a whole. Substituting these values in equation (2.1), the estimated probability density function of the age at marriage in Andhra Pradesh is

$$\hat{f}(x) = (0.00012)(x-13)^{4.3} e^{-0.00006(x-13)^{5.3}}$$

$$+(0.202)(x-13)^{2.98} e^{-0.083(x-13)^{3.98}}, x > 13$$

From this probability distribution, various characteristics of the variable can be obtained. The estimated mean age at marriage in Andhra Pradesh is 16.546 years with the variability 4.135. The minimum age at marriage is estimated as 13 years and maximum age at marriage is estimated as 33 years.

The probability that women up to the age at '*t*' will remain unmarried in Andhra Pradesh is

$$\bar{F}(t) = 1 - F(t) = 1 - \int_{13}^{t} f(x)dx$$

$$= 1 - \int_{15}^{i} \left(p\frac{c_1}{a_1}\left(\frac{x-b}{a_1}\right)^{c_1-1} e^{-\left(\frac{x-b}{a_1}\right)^{c_1}} + q\frac{c_2}{a_2}\left(\frac{x-b}{a_2}\right)^{c_2-1} e^{-\left(\frac{x-b}{a_2}\right)^{c_2}}\right) dx.$$

## *2.1 Regionwise Distribution of Age at Marriage in Andhra Pradesh*

Andhra Pradesh is composed of three major regions, namely, Coastal Andhra, Rayalaseema and Telangana. These three regions varied over different social and cultural activities whose impact on age at marriage is of greater importance. Hence to have the planning at state level, a micro level regional planning is required. Accordingly, it is needed to develop probability distributions for age at marriage in each region of Andhra Pradesh.

For Coastal Andhra region, a sample of 2,197 married women is collected and presented in Table 7.2.

**Table 7.2 :** Bivariate data on Age at Marriage and Age at Sterilization in Costal Andhra Region

| X \ Y | 16 - 19 | 19 - 22 | 22 - 25 | 25 - 28 | 28 - 31 | 31 - 34 | 34 - 37 | Total |
|---|---|---|---|---|---|---|---|---|
| 13 – 16 | 83 | 482 | 318 | 185 | 67 | - | - | 1,135 |
| 16 – 19 | - | 112 | 441 | 248 | 36 | - | - | 837 |
| 19 – 22 | - | - | 85 | 64 | 23 | 3 | - | 175 |
| 22 – 25 | - | - | - | 23 | 12 | 7 | - | 42 |
| 25 – 28 | - | - | - | - | 1 | 6 | - | 7 |
| 28 – 31 | - | - | - | - | - | - | 1 | 1 |
| 31 – 34 | - | - | - | - | - | - | - | - |
| Total | 83 | 594 | 844 | 520 | 139 | 16 | 1 | 2,197 |

From the frequency curve of this data, it is observed that it is bimodal having one mode at 15 years another at 18 years and justifying a left truncated mixture of Weibull distribution.

The maximum likelihood estimates of the parameters are, $p = 0.393$, $q = 0.607$, $a_1 = 6.227$, $c_1 = 5.5$, $a_2 = 2.4$, $c_2 = 3.98$ and $b = 13$. Substituting these values in equation (2.1), the estimated probability density function of the age at marriage in Coastal Andhra region is,

$$\hat{f}(x) = (0.00009)(x-13)^{4.5} e^{-0.00004(x-13)^{5.5}}$$

$$+(0.074)(x-13)^{2.98} e^{-0.03(x-13)^{3.98}}, \; x > 13$$

The estimated mean age at marriage in Coastal Andhra

region is 16.55 years with the variability 4.303.

For Rayalaseema region, from a sample of 943 married women is collected and presented in Table 7.3.

**Table 7.3 :** Bivariate data on Age at Marriage and Age at Sterilization in Rayalaseema region

| X \ Y | 16 - 19 | 19 - 22 | 22 - 25 | 25 - 28 | 28 - 31 | 31 - 34 | 34 - 37 | Total |
|---|---|---|---|---|---|---|---|---|
| 13 – 16 | 13 | 131 | 128 | 115 | 96 | - | - | 483 |
| 16 – 19 | - | 5 | 173 | 151 | 5 | - | - | 334 |
| 19 – 22 | - | - | 25 | 55 | 11 | 11 | - | 102 |
| 22 – 25 | - | - | - | 8 | 5 | 4 | - | 17 |
| 25 – 28 | - | - | - | - | 1 | 3 | - | 4 |
| 28 – 31 | - | - | - | - | - | 2 | 1 | 3 |
| 31 – 34 | - | - | - | - | - | - | - | - |
| **Total** | 13 | 136 | 326 | 329 | 118 | 20 | 1 | 943 |

From the frequency curve of this data it is also observed to be a bimodal and justifying a mixture of left truncated Weibull distribution. The estimated values of the parameters of the model are $p = 0.393$, $q = 0.607$, $a_1 = 6.227, c_1 = 5.5$, $a_2 = 2.4$, $c_2 = 2.98$ and $b = 13$. Substituting these values in equation (2.1), the estimated probability density function of the age at marriage in Rayalaseema region is obtained as

$$\hat{f}(x) = (0.00009)(x-13)^{4.5} e^{-0.00004(x-13)^{5.5}}$$

$$+(0.133)(x-13)^{1.98} e^{-0.074(x-13)^{2.98}}, x > 13.$$

The estimated mean age of marriage in Rayalaseema region is 16.56 years with variability 4.047 years.

For Telangana region, a sample of 2,072 married women is collected and presented in Table 7.4.

From the frequency curve of this data, it is also observed that a mixture of left truncated Weibull distribution fits the data. The estimated values of the parameters of the model are $p = 0.393, q = 0.607, a_1 = 6, c_1 = 5.2, a_2 = 2.4, c_2 = 3.9$ and $b = 13$. Substituting these values in equation (2.1), the estimated probability density function of the age at marriage in

**Table 7.4** : Bivariate data on Age at Marriage and Age at Sterilization in Telangana region

| X \ Y | 16 - 19 | 19 - 22 | 22 - 25 | 25 - 28 | 28 - 31 | 31 - 34 | 34 - 37 | Total |
|---|---|---|---|---|---|---|---|---|
| 13 – 16 | 43 | 375 | 313 | 301 | 71 | 5 | - | 1,108 |
| 16 – 19 | - | 40 | 337 | 258 | 100 | 6 | - | 741 |
| 19 – 22 | - | - | 115 | 20 | 15 | 13 | - | 163 |
| 22 – 25 | - | - | - | 27 | 11 | 3 | - | 41 |
| 25 – 28 | - | - | - | - | 10 | 5 | 1 | 16 |
| 28 – 31 | - | - | - | - | - | 1 | 1 | 2 |
| 31 – 34 | - | - | - | - | - | - | 1 | 1 |
| **Total** | 13 | 415 | 765 | 606 | 207 | 33 | 3 | 2,072 |

Telangana region is obtained as,

$$\hat{f}(x) = (0.00018)(x-13)^{4.2} e^{-0.00009(x-13)^{5.2}}$$

$$+(0.078)(x-13)^{2.9} e^{-0.033(x-13)^{3.9}}, x > 13.$$

The estimated mean age of marriage in Telangana region is 16.488 years with variability 3.497.

### 3. Probability Distribution of Age at Sterilization

Contraception is a major factor, which influences the fertility of a population. The terminal methods of Family Planning, which are permanent in nature and technically known as Tubectomy for females and Vasectomy for males have significant impact on present day population dynamics. Several authors have studied the role of sterilization in controlling population growth. In populations where birth control through spacing methods is least effective, sterilization plays a dominant role. Also the female sterilization which is more popular and widely practised in societies like the study area (Andhra Pradesh) it is the single most Family Planning method which accounts for about 95 per cent of the total contraception. There is limited adoption of vasectomy.

However, the adoption of the method by one of the

partners of the couple is sufficient enough for limiting the family size. Because of small family norm advocated by the government through their machinery, the age at sterilization of women is drastically reduced during the last two decades. It is also observed through different demographic studies that the age at marriage has a tremendous influence on age at sterilization. It is also a fact that the age at sterilization is directly related to the health of women, since this sterilization avoids the forcible MTP (Medical Termination of Pregnancy). So, the pattern of age at sterilization is also useful along with age at marriage, to have the idea about fertility pattern of a region. Since the age at sterilization got influenced by various socio-economic factors, which are random in nature, the age at sterilization can be taken as random. The reliability concepts can also be applied to age at sterilization and can be taken as equivalent to that of failure time, as sterilization can be considered a failure for giving birth to a child. The random nature of age at sterilization can be studied by ascribing a suitable probability distribution to the age at sterilization. So from the data collected in RCH survey, the frequency distribution of age at sterilization of women from the sample of 5,212 in Andhra Pradesh is obtained and shown in Table 7.1.

From the frequency curve of age at sterilization, it is observed that a left truncated Weibull distribution having the probability density function of the form

$$g(y)=\frac{m}{k}\left(\frac{y-n}{k}\right)^{m-1} e^{-\left(\frac{y-n}{k}\right)^{m}}, \; y > n$$

is a suitable probability function of age at sterilization, since the curve shows that there is a long upper tail. The estimates of the parameters for Andhra Pradesh are $m = 2.959, k = 9.112, n = 17$.

The estimated probability density function of the age at sterilization for Andhra Pradesh is,

$$\hat{g}(y) = (0.0043)(y-17)^{1.959}\, e^{-0.00145(y-17)^{2.959}}, \quad y>17.$$

The mean age of sterilization of a woman is 25.132 years

with variability 8.952.

For the Coastal Andhra region, a sample of 2,197 sterilized women is collected and presented in Table 7.2.

From frequency curve of this data it is also to be observed that a left truncated Weibull distribution is a suitable probability distribution for the age at sterilization of women in Coastal Andhra region of Andhra Pradesh.

The estimated probability density function of age at sterilization of women in Coastal Andhra region is obtained as,

$$\hat{g}(y) = (0.0155)(y-17)^{1.542}\, e^{-0.00608(y-17)^{2.542}}, \; y > 17.$$

The mean age of sterilization of women is 23.605 years with variability 7.751.

For the Rayalaseema region, a sample of 943 sterilized women is collected and presented in Table 7.3.

From the frequency curve of this data, it is also observed that the left truncated Weibull distribution is a suitable probability distribution for the age at sterilization of women in Rayalaseema region.

The estimated probability density function of age at sterilization of women in Rayalaseema region is,

$$\hat{g}(y) = (0.0033)(y-17)^{2.134}\, e^{-0.00105(y-17)^{3.134}}, \; y > 17.$$

The mean age at sterilization of women is 24.975 years with variability 7.768 years.

For the Telangana region, a sample of 2,072 sterilized women is collected and presented in Table 7.4.

From the frequency curve of this data, it is also observed that the left truncated Weibull distribution is a suitable probability distribution for the age at sterilization of women in Telangana region.

The estimated probability density function of age at sterilization of women in Telangana region is,

$$\hat{g}(y) = (0.0123)(y-17)^{1.503} e^{-0.00492(y-17)^{2.503}}, \; y > 17.$$

T!✱ mean age at sterilization of women is 24.413 years with variability 10.056 years.

The probability that a woman will be sterilized before attaining the age '*t*' is

$$h(t) = \int_{17}^{t} g(t)dt \, .$$

For different values of t, the values of $\hat{h}(t)$ foı ʌndhra Pradesh, Coastal Andhra, Telangana and Rayalaseema are computed and presented in Table 7.5.

**Table 7. 5 :** Values of $\hat{h}(t)$

| Age (t) | $\hat{h}(t)$ | | | |
|---|---|---|---|---|
| | *Andhra Pradesh State* | *Coastal Andhra Region* | *Rayalaseema Region* | *Telangana Region* |
| 17 | 0 | 0 | 0 | 0 |
| 18 | 0.001 | 0.005 | 0.001 | 0.001 |
| 19 | 0.011 | 0.028 | 0.011 | 0.009 |
| 20 | 0.026 | 0.075 | 0.026 | 0.032 |
| 21 | 0.073 | 0.147 | 0.073 | 0.032 |
| 22 | 0.145 | 0.242 | 0.145 | 0.078 |
| 23 | 0.241 | 0.354 | 0.241 | 0.151 |
| 24 | 0.357 | 0.474 | 0.357 | 0.251 |
| 25 | 0.483 | 0.592 | 0.483 | 0.374 |
| 26 | 0.608 | 0.700 | 0.608 | 0.509 |
| 27 | 0.721 | 0.791 | 0.721 | 0.643 |
| 28 | 0.814 | 0.863 | 0.814 | 0.761 |
| 29 | 0.884 | 0.915 | 0.884 | 0.855 |
| 30 | 0.931 | 0.951 | 0.931 | 0.921 |
| 31 | 0.960 | 0.973 | 0.960 | 0.962 |
| 32 | 0.976 | 0.986 | 0.976 | 0.984 |
| 33 | 0.984 | 0.993 | 0.984 | 0.994 |
| 34 | 0.987 | 0.996 | 0.987 | 0.998 |
| 35 | 0.988 | 0.998 | 0.988 | 0.999 |
| 36 | 0.988 | 0.998 | 0.988 | 0.999 |
| 37 | 0.989 | 0.999 | 0.989 | 0.999 |

## 4. Joint Distribution of Age at Marriage and Age at Sterilization

Let (X,Y) be a bivariate random variable with marginal probability density function f(x) of the variable 'X', and marginal probability density function g(y) of the variable 'Y', $F(x)$ be the probability distribution function of variable 'X', $G(y)$ be the probability distribution function of variable 'Y' and $\theta$ be the combined parameter. Then the Morgenstern's form (Nevett,1967) of the joint probability density function of (X,Y) is

$$f(x,y) = f(x)g(y)[1+\theta\{2F(x)-1\}\{2G(y)-1\}] \qquad \dots (4.1)$$

For finding the joint probability density function of age at marriage (X) and age at Sterilization (Y), the equation (4.1) was utilized. From sections 2 and 3 the probability density function of age at marriage is

$$f(x) = p\frac{c_1}{a_1}\left(\frac{x-b}{a_1}\right)^{c_1-1} e^{-\left(\frac{x-b}{a_1}\right)^{c_1}} + q\frac{c_2}{a_2}\left(\frac{x-b}{a_2}\right)^{c_2-1} e^{-\left(\frac{x-b}{a_2}\right)^{c_2}},$$

$$x > b, \qquad \dots (4.2)$$

and the probability density function of age at sterilization is

$$g(y) = \frac{m}{k}\left(\frac{y-x}{k}\right)^{m} e^{-\left(\frac{y-x}{k}\right)^{m}}, \quad y>x \qquad \dots (4.3)$$

Then,

$$F(x) = p\left[1-e^{-\left(\frac{x-b}{a_1}\right)^{c_1}}\right] + q\left[1-e^{-\left(\frac{x-b}{a_2}\right)^{c_2}}\right]$$

$$= 1 - pe^{-\left(\frac{x-b}{a_1}\right)^{c_1}} - qe^{-\left(\frac{x-b}{a_2}\right)^{c_2}} \qquad \dots (4.4)$$

$$G(y) = 1 - e^{-\left(\frac{y-x}{k}\right)^m} \qquad \dots (4.5)$$

Substituting the equations (4.2) to (4.5) in equation (4.1), we get $f(x, y)$

$$= \left[ p\frac{c_1}{a_1}\left(\frac{x-b}{a_1}\right)^{c_1-1} e^{-\left(\frac{x-b}{a_1}\right)^{c_1}} + q\frac{c_2}{a_2}\left(\frac{x-b}{a_2}\right)^{c_2-1} e^{-\left(\frac{x-b}{a_2}\right)^{c_2}} \right]\left[\frac{m}{k}\left(\frac{y-x}{k}\right)^{m-1} e^{-\left(\frac{y-x}{k}\right)^m}\right]$$

$$\left[1+\theta\left(1-2pe^{-\left(\frac{x-b}{a_1}\right)^{c_1}} - 2qe^{-\left(\frac{x-b}{a_2}\right)^{c_2}}\right)\left(1-2e^{-\left(\frac{y-x}{k}\right)^m}\right)\right]$$

$$= \left[ p\frac{c_1}{a_1}\left(\frac{x-b}{a_1}\right)^{c_1-1} e^{-\left(\frac{x-b}{a_1}\right)^{c_1}} + q\frac{c_2}{a_2}\left(\frac{x-b}{a_2}\right) e^{-\left(\frac{x-b}{a_2}\right)^{c_2}} \right]$$

$$\times \left[\left\{1+\theta - 2p\theta e^{-\left(\frac{x-b}{a_1}\right)^{c_1}} - 2q\theta e^{-\left(\frac{x-b}{a_2}\right)^{c_2}}\right\}\left\{\frac{m}{k}\left(\frac{y-x}{k}\right)^{m-1} e^{-\left(\frac{y-x}{k}\right)^m}\right\}\right.$$

$$\left. + \left\{4p\theta e^{-\left(\frac{x-b}{a_1}\right)^{c_1}} + 4q\theta e^{-\left(\frac{x-b}{a_2}\right)^{c_2}} - 2\theta\right\}\left\{\frac{m}{k}\left(\frac{y-x}{k}\right)^{m-1} e^{-2\left(\frac{y-x}{k}\right)^m}\right\}\right],$$

$$x > b, y > x.$$

## 5. Distributional Properties

The mean vector of the joint distribution is

$$\begin{bmatrix} E(X) \\ E(Y) \end{bmatrix} = \begin{bmatrix} p\left[b + a_1\Gamma\left(\frac{1}{c_1}+1\right)\right] + q\left[b + a_2\Gamma\left(\frac{1}{c_2}+1\right)\right] \\ n + k\Gamma\left(\frac{1}{m}+1\right) \end{bmatrix}$$

The covariance between age at marriage and age at sterilization is

$$Cov(X,Y) = E(XY) - E(X)E(Y),$$

where $E(XY) = \int_b^{\infty}\int_x^{\infty} xy f(x,y)\,dx\,dy$.

After substituting the value for E(XY) we get,

$$Cov(X,Y) = p\left[b^2 + 2ba_1\Gamma\left(\frac{1}{c_1}+1\right) + a_1^2\Gamma\left(\frac{2}{c_1}+1\right)\right]$$

$$+p\,k\Gamma\left(\frac{1}{m}+1\right)\left[1+\theta\left(1-2^{-\frac{1}{m}}\right)\right]\left[b+a_1\Gamma\left(\frac{1}{c_1}+1\right)\right]$$

$$-p^2\theta\,k\Gamma\left(\frac{1}{m}+1\right)\left(1-2^{-\frac{1}{m}}\right)\left[b+2^{-\frac{1}{c_1}}a_1\Gamma\left(\frac{1}{c_1}+1\right)\right]$$

$$-2pq\,\theta\,k\Gamma\left(\frac{1}{m}+1\right)\left(1-2^{-\frac{1}{m}}\right)\int_b^{\infty} x\frac{c_1}{a_1}\left(\frac{x-b}{a_1}\right)^{c_1-1} e^{-\left(\frac{x-b}{a_2}\right)^{c_2}} e^{-\left(\frac{x-b}{a_1}\right)^{c_1}}\,dx$$

$$+q\left[b^2 + 2ba_2\Gamma\left(\frac{1}{c_2}+1\right) + a_2^2\Gamma\left(\frac{2}{c_2}+1\right)\right]$$

$$+q\,k\Gamma\left(\frac{1}{m}+1\right)\left[1+\theta\left(1-2^{-\frac{1}{m}}\right)\right]\left[b+a_2\Gamma\left(\frac{1}{c_2}+1\right)\right]$$

$$-2pq\,\theta\,k\Gamma\left(\frac{1}{m}+1\right)\left(1-2^{-\frac{1}{m}}\right)\int_b^{\infty} x\frac{c_2}{a_2}\left(\frac{x-b}{a_2}\right)^{c_2-1} e^{-\left(\frac{x-b}{a_2}\right)^{c_2}} e^{-\left(\frac{x-b}{a_1}\right)^{c_1}}\,dx$$

$$-q^2\theta\ k\Gamma\left(\frac{1}{m}+1\right)\left(1-2^{-\frac{1}{m}}\right)\left[b+2^{-\frac{1}{c_2}}a_2\Gamma\left(\frac{1}{c_2}+1\right)\right]$$

$$-\left[p\left[b+a_1\Gamma\left(\frac{1}{c_1}+1\right)\right]+q\left[b+a_2\Gamma\left(\frac{1}{c_2}+1\right)\right]\left[n+k\Gamma\left(\frac{1}{m}+1\right)\right]\right] \quad \ldots(5.1)$$

The variance covariance matrix of the joint distribution is

$$\Sigma=\begin{bmatrix} V(X) & Cov(X,Y) \\ Cov(X,Y) & V(Y) \end{bmatrix},$$

where

$$V(X)=p\left[b^2+2ba_1\Gamma\left(\frac{1}{c_1}+1\right)+a_1{}^2\Gamma\left(\frac{2}{c_1}+1\right)\right]+$$

$$\left[b^2+2ba_1\Gamma\left(\frac{1}{c_1}+1\right)+a_1{}^2\Gamma\left(\frac{2}{c_1}+1\right)\right]$$

$$-\left[p\left(b+a_1\Gamma\left(\frac{1}{c_1}+1\right)\right)+q\left(b+a_2\Gamma\left(\frac{1}{c_2}+1\right)\right)\right]^2$$

$$V(Y)=k^2\left[\Gamma\left(\frac{2}{m}+1\right)-\left(\Gamma\left(\frac{1}{m}+1\right)\right)^2\right]$$

and $Cov(X,Y)$ is as in equation (5.1).

The correlation coefficient between age at marriage and age at sterilization is

$$r(X,Y) = \frac{Cov(X,Y)}{[V(X)V(Y)]^{\frac{1}{2}}}$$

$$= \frac{z_1 + z_2 - z_3 - z_4 + z_5 + z_6 - z_7 - z_8 - z_9}{(z_{10} z_{11})^{\frac{1}{2}}},$$

where

$$z_1 = p\left[b^2 + 2ba_1\Gamma\left(\frac{1}{c_1}+1\right) + a_1^{\,2}\Gamma\left(\frac{2}{c_1}+1\right)\right]$$

$$z_2 = p\; k\Gamma\left(\frac{1}{m}+1\right)\left[1+\theta\left(1-2^{-\frac{1}{m}}\right)\right]\left[b + a_1\Gamma\left(\frac{1}{c_1}+1\right)\right]$$

$$z_3 = p^2\theta\; k\Gamma\left(\frac{1}{m}+1\right)\left(1-2^{-\frac{1}{m}}\right)\left[b + 2^{-\frac{1}{c_1}} a_1\Gamma\left(\frac{1}{c_1}+1\right)\right]$$

$$z_4 = 2pq\,\theta\; k\Gamma\left(\frac{1}{m}+1\right)\left(1-2^{-\frac{1}{m}}\right)\int_b^\infty x\frac{c_1}{a_1}\left(\frac{x-b}{a_1}\right)^{c_1-1} e^{-\left(\frac{x-b}{a_2}\right)^{c_2}} e^{-\left(\frac{x-b}{a_1}\right)^{c_1}} dx$$

$$z_5 = q\left[b^2 + 2ba_2\Gamma\left(\frac{1}{c_2}+1\right) + a_2^{\,2}\Gamma\left(\frac{2}{c_2}+1\right)\right]$$

$$z_6 = q\, k\Gamma\left(\frac{1}{m}+1\right)\left[1+\theta\left(1-2^{-\frac{1}{m}}\right)\right]\left[b + a_2\Gamma\left(\frac{1}{c_2}+1\right)\right]$$

$$z_7 = 2pq\,\theta\; k\Gamma\left(\frac{1}{m}+1\right)\left(1-2^{-\frac{1}{m}}\right)\int_b^\infty x\frac{c_2}{a_2}\left(\frac{x-b}{a_2}\right)^{c_2-1} e^{-\left(\frac{x-b}{a_2}\right)^{c_2}} e^{-\left(\frac{x-b}{a_1}\right)^{c_1}} dx$$

$$z_8 = q^2\theta\; k\Gamma\left(\frac{1}{m}+1\right)\left(1-2^{-\frac{1}{m}}\right)\left[b + 2^{-\frac{1}{c_2}} a_2\Gamma\left(\frac{1}{c_2}+1\right)\right]$$

$$z_9 = \left[ p\left(b + a_1\Gamma\left(\frac{1}{c_1}+1\right)\right) + q\left(b + a_2\Gamma\left(\frac{1}{c_2}+1\right)\right)\right]\left[n + k\Gamma\left(\frac{1}{m}+1\right)\right]$$

$$z_{10} = p\left[b^2 + 2ba_1\Gamma\left(\frac{1}{c_1}+1\right) + a_1{}^2\Gamma\left(\frac{2}{c_1}+1\right)\right]$$

$$+ \; q\left[b^2 + 2ba_1\Gamma\left(\frac{1}{c_1}+1\right) + a_1{}^2\Gamma\left(\frac{2}{c_1}+1\right)\right]$$

$$- \left[ p\left(b + a_1\Gamma\left(\frac{1}{c_1}+1\right)\right) + q\left(b + a_2\Gamma\left(\frac{1}{c_2}+1\right)\right)\right]^2$$

$$z_{11} = k^2\left[\Gamma\left(\frac{2}{m}+1\right) - \left(\Gamma\left(\frac{1}{m}+1\right)\right)^2\right].$$

The conditional probability density function *of Y, given X=x, is*

$$h(x/y) = \frac{f(x,y)}{g(x)}$$

$$= \left[1 + \theta - 2p\theta e^{-\left(\frac{x-b}{a_1}\right)^{c1}} - 2q\theta e^{-\left(\frac{x-b}{a2}\right)^{c2}}\right]\left[\frac{m}{k}\left(\frac{y-x}{k}\right)^{m-1} e^{-\left(\frac{y-x}{k}\right)^m}\right]$$

$$+ \left[4p\theta e^{-\left(\frac{x-b}{a_1}\right)^{c1}} - 4q\theta e^{-\left(\frac{x-b}{a2}\right)^{c2}} - 2\theta\right]\left[\frac{m}{k}\left(\frac{y-x}{k}\right)^{m-1} e^{-2\left(\frac{y-x}{k}\right)^m}\right]$$

The conditional probability density function *of X, given Y=y, is*

$$h(x/y) = \frac{f(x,y)}{g(y)}$$

$$= \left[p\frac{c_1}{a_1}\left(\frac{x-b}{a_1}\right)^{c_1-1} e^{-\left(\frac{x-b}{a_1}\right)^{c_1}} + q\frac{c_2}{a_2}\left(\frac{x-b}{a_2}\right) e^{-\left(\frac{x-b}{a_2}\right)^{c_2}}\right] \times$$

$$\left[1+\theta-2\theta e^{-\left(\frac{y-x}{k}\right)^{m}}-2p\theta e^{-\left(\frac{x-b}{a_1}\right)^{c_1}}+4p\theta e^{-\left(\frac{x-b}{a_1}\right)^{c_1}}e^{-\left(\frac{y-x}{k}\right)^{m}}-2q\theta e^{-\left(\frac{x-b}{a_2}\right)^{c_2}}\right.$$

$$\left.+4q\theta e^{-\left(\frac{x-b}{a_2}\right)^{c_2}}e^{-\left(\frac{y-x}{k}\right)^{m}}\right].$$

## 6. Regression Model for Age at Sterilization on Age at Marriage

To study the effect of age at marriage on age at sterilization, the conditional expectation of age at sterilization given the value of age at marriage was obtained. Using the conditional expectation of age at sterilization on age at marriage is given by Umakrishna (2002) as

$$E(Y/X=x)=\int_x^{\infty} y\ h(y/x)\,dy$$

$$=\left[1+\theta-2p\theta e^{-\left(\frac{x-b}{a_1}\right)^{c_1}}-2q\theta e^{-\left(\frac{x-b}{a_2}\right)^{c_2}}\right]\int_x^{\infty} y\frac{m}{k}\left(\frac{y-x}{k}\right)^{m-1}e^{-\left(\frac{y-x}{k}\right)^{m}}dy$$

$$+\left[4p\theta e^{-\left(\frac{x-b}{a_1}\right)^{c_1}}+4q\theta e^{-\left(\frac{x-b}{a_2}\right)^{c_2}}-2\theta\right]\int_x^{\infty} y\frac{m}{k}\left(\frac{y-x}{k}\right)^{m-1}e^{-2\left(\frac{y-x}{k}\right)^{m}}dy$$

$$E(Y/X=x)$$

$$=x+k\Gamma\left(\frac{1}{m}+1\right)+\theta\left(1-2pe^{-\left(\frac{x-b}{a_1}\right)^{c_1}}\right)\left(-2pe^{-\left(\frac{x-b}{a_2}\right)^{c_2}}\right)$$

$$\left(k\Gamma\left(\frac{1}{m}+1\right)\left(1-2^{-\frac{1}{m}}\right)\right). \qquad \text{... (6.1)}$$

The equation (6.1) represents a curve. The regression curve of age at sterilization on age at marriage is given by,

$$Y = E(Y/X = x) + \varepsilon, \text{ where } \varepsilon \text{ is the residual.} \quad \dots (6.2)$$

Substituting the equation (6.1) in the equation (6.2), one can get the regression curve as

$$Y = x + k\Gamma\left(\frac{1}{m}+1\right) + \theta\left(1 - 2pe^{-\left(\frac{x-b}{a_1}\right)^{c_1}} - 2pe^{-\left(\frac{x-b}{a_2}\right)^{c_2}}\right)$$

$$\left(k\Gamma\left(\frac{1}{m}+1\right)\left(1-2^{-\frac{1}{m}}\right)\right) + \varepsilon.$$

The parameters $p, q, a_1, a_2, b, c_1, c_2, k$ and $m$ can be estimated using the method of moments by equating the sample moments with the theoretical moments. The combined parameter $\theta$ can be estimated by equating the sample correlation coefficient with theoretical correlation coefficient.

## 7. Conclusion

From the probability distributions of age at marriage, it is found that the mean age at marriage in Andhra Pradesh is 16.54 years with variability 4.135 years. Among the three regions of Andhra Pradesh, no difference in mean age at marriage was obtained and naturally same as that of the State of Andhra Pradesh. But the variability in age at marriage for Coastal Andhra region is higher when compared to other two regions, out of which Telangana region stood at low level. The maximum age at marriage for entire Andhra Pradesh is found to be 33 years. Regarding maximum age at marriage in three regions of Andhra Pradesh, Coastal Andhra region has the lowest age when compared to other two regions. The age at marriage for the entire state and for the three regions of Andhra Pradesh follow a skewed bimodal distribution.

From the probability distribution of age at sterilization, the mean age at sterilization in the entire Andhra Pradesh State is 25.132 years with variability 8.952 years. Regarding Coastal Andhra region, the mean age at sterilization is lower compared to the remaining two regions and Andhra Pradesh as a whole. The variability in age at sterilization for Telangana region is much higher than the remaining two regions and Andhra Pradesh as a whole. The maximum age at sterilization in Andhra Pradesh is found to be 36 years.

Table 7.5 provides the probability that a woman is sterilized before reaching a specified age. It is also observed that the probabilities go on increasing corresponding to the increase of ages in all the three regions of Andhra Pradesh and Andhra Pradesh State as a whole. Regarding Coastal Andhra and Rayalaseema regions the marriages are performed at an early stage when compared to Telangana region and Andhra Pradesh as a whole. These distributions are useful to develop regression surfaces of age at sterilization on age at marriage, by using joint probability density function of these two variables and conditional expectation.

## References


Anderson, K.H., Hill, M.A. and Butler, J.S. (1987), Age at Marriage in Malaysia : a Hazard Model of Marriage Timing, *Journal of Development Economics*, 26, 223-234.

Audinarayana, N. (1993), Determinants of Female Age at Marriage in Rural Andhra Pradesh : Multivariate Analysis, *Demograpy India*, 22(2), 169- 740.

Glick, P.C. (1967), Marriage and Family Variables Related to Fertility, *Proceedings of the World Population Conference, Belgrade, 30 August-10 September 1965, 2. Selected Papers and Summaries: Fertility, Family Planning, Mortality, E/CONF.* 41/3, 210-213.

Goyal, R.P. (1989), Raising Female Age at Marriage : Priority in India's Population Policy and Programme, *Demography India*, 18, 177-182.

Husain, I.Z. (1971), Mean Age at Marriage and Differential Fertility, *International Union for Scientific Study of Population (IUSSP), International Population Conference, London*, 1, 466.

Johnson, and Kotz (1974), *Continuous Univariate Distributions*, John Wiley and Sons.

Karkal, M. and Rajan, S.I. (1989), Age at Marriage : How Much Change, *Economic and Political Weekly*, 24, 505-506.

Murty, V. M. (2001), *Andhra Pradesh RHS-RCH Project Report*, Population Research Centre, Andhra University, Visakhapatnam.

Nair, P.S., Koteshwar, R.K. (1987), Female Age at Marriage in Northern Karnataka, *Social Change*, 17, 65-70.

Nath, D.C., Land, K.C. and Talukdar, P.K. (1992), A Model that Fits Female Age at First Marriage in a Traditional Society, *Janasamkhya*, 10, 53-59

Nevett, A. (1967), Age at Marriage, Parental Responsibility and the Size of the Family, *Proceedings of the World Population Conference, Belgrade, 30 August – 10 September 1965* . 2. *Selected Papers and Summaries* : *Fertility, Family Planning, Mortality. E/CONF*. 41/3, 176-177.

Ramabhadran, V.K. (1987), *Monograph on Age at Marriage in India*, New Delhi, India, Family Planning Foundation.

Sinha, R.K. (1987), Impact of Age at Marriage on Fertility and Completed Family Size in Eastern Rajasthan, *Journal of Family Welfare*, 34, 32-40.

Umakrishna, K. (2002), *Some Stochastic Models for Age at Marriage and Age at Sterilization*, Thesis submitted in Andhra University, Visakhapatnam.

Wang, F. and Vang, Q. (1996), Age at Marriage and the First Birth Interval : The Emerging Change in Sexual Behaviour among Young Couples in China, *Population and Development Review*, 22, 299-320.

8

# Stochastic Models for Unemployment in a Region

*K. Srinivasa Rao*

## 1. Introduction

At present there are about 5 million entrants on the employment register every year in India. Some of the entrants leave the employment register within a few months and others remain on the rolls of employment register for any time up to five years or more. There are, therefore, enormous variation in the duration of spells of unemployment from one individual to the other because of various factors like industrialization, qualifications, economic conditions, wage structure, Government policies etc., which are random in nature. The information on the unemployment spells of individuals in any region are needed in order to analyze the conditions of the people therein and to develop suitable welfare or developmental programmes. The information on unemployment duration of individuals also provides a significant impact on rural and industrial development. Several researchers investigated the phenomenon of unemployment. Uzuwa (1964) pioneered the systematic study of unemployment growth. Oliver (1964) observed that high relative regional unemployment is associated with high net migration. Horowitz (1970) studied the employment concentration in common market through an entropy approach. Richard and Barrett (1974) investigated the retrospective bias in unemployment reported by sex, race and age. Cargill and Meyer (1974) studied the prices and unemployment using distributed lag models. Baily (1977) developed the theory of lay offs in analyzing the labour force. Maria and Christinehoza (1978) studied the small area estimation with application to unemployment. Frank (1978) obtained the distribution of completed unemployment spell lengths. Caster (1979) proposed a parametric function for duration of unemployment distribution. Caster and Nickell (1980) considered the modelling and estimation of the process covering the rate at which unemployed individuals leave on unemployment. Aitkin and Healey (1985) modelled the unemployment rates from the EEC labour force survey. Moffit (1985) studied the unemployment insurance and distribution of unemployment spells using hazard plots. Casler (1988) studied the unemployment and unemployment duration in Spain. Meyer (1990) studied the effect of unemployment benefit on employment duration. Storer (1994) studied the unemployment dynamics and labour market. Korpi (1995) studied the effects of manpower policies in duration on re-employment rates. James (1995) calculated the distribution of the unemployment duration length and transition rates. However, a very little work has been done regarding the stochastic modeling of unemployment and its duration except the contributions based on search theory and unemployment insurance benefits. Hence in this paper we develop and analyze some stochastic models for number of unemployed persons in a region and probability distribution of unemployment duration using stochastic processes.

## 2. Stochastic Model for the Number of Unemployed Persons in an Employment Exchange

Let the employment process in an exchange be Poisson with parameter $a$. Let the recruitment process also be $a$ Poisson with parameter $b$. Let us assume that all the recruitments in the region are through employment exchange. Let $P_n(t)$ be the probability that there are $n$ individuals enrolled and remaining unemployed in the employment exchange at time $t$. The difference-differential equations of the model are

$$dP_n(t)/dt = -(a+b)P_n(t) + bP_{n+1}(t) + aP_{n-1}(t), \text{for } n \geq 1$$

$$dP_0(t)/dt = -aP_0(t) + bP_1(t). \qquad \text{... (2.1)}$$

Assuming that the process of unemployment is stable, the steady-state equations of the model(using 2.1) are

$$-(a+b)P_n + bP_{n+1} + aP_{n-1} = 0, \text{for } n \geq 1$$

$$-aP_0 + bP_1 = 0 \qquad \text{... (2.2)}$$

Solving the equation (2.2), the probability mass function of the number of unemployed individuals in the employment exchange is

$$P_n = [(b-a)/b][a/b]^n, \quad \text{for } n = 0.1, 2, \ldots$$

The average number of unemployed persons in the employment exchange is

$$N = \frac{a}{(ba- )}$$

The probability that there is no unemployed person in the employment exchange is

$$[(b-a)/b].$$

The variability of the number of unemployed persons in the employment exchange is

$$V = \frac{ab}{(b-a)^2}.$$

The coefficient of variation is $C.V. = \sqrt{(b/a)}$.

### 3. Distribution of Unemployment Duration

Let $T$ be the random variable of the unemployment period

of an individual registered in the employment exchange, and $U(t)$ be the cumulative distribution function of $T$. Then we have

$$U(0) = P[T \le 0] = P(T = 0) = (b - a)/b.$$

Consider $U(t)$, the probability of an individual having unemployment period less than or equal to $t$. If there are $n$ individuals already registered in the exchange, upon the registration of the $(n+1)^{st}$ individual to get a job at a time period $(0, t)$, all prior registered individuals must have been absorbed by time $t$. Since the recruitment process is memory less, the distribution of the time required for $n$ individuals recruitment is independent of the time of the $(n+1)^{st}$ individual registration and is the convolution of $n$ exponential random variables which is of Erlang type $n$. In addition, since the enrollment is Poisson, the probability that an individual who registered in the employment exchange finds $n$ already registered individuals in the system is identical to the equilibrium distribution of the number of unemployed enrolled individuals.

Then $U(t) = P(T \le t) = U(0) + \sum_{n-1}^{\infty} \Pr[(n-1)$ *persons are recruited at time t | there are n individuals registered for jobs* in the employment exchange]

$$= \sum_{n=1}^{\infty}\left[1-\frac{a}{b}\right]\left(\frac{a}{b}\right)^{n}\int_{0}^{t}\frac{(bx)^{n-1}}{(n-1)!}e^{-bx}b\,dx + \left[1-\frac{a}{b}\right]$$

$$= \left(1-\frac{a}{b}\right)\left(\frac{a}{b}\right)\int_{0}^{t} be^{-bt\left(1-\frac{a}{b}\right)}dt + \left(1-\frac{a}{b}\right)$$

$= 1-(a/b)\, e^{-(b-a)t}$, for $t>0$.

The expected duration of unemployment of an individual registered in the employment exchange is $U = a/[b(b-a)]$. The variability of the unemployment period is

$$V = [2ab - a^2]/[b^2(b-a)^2].$$

The coefficient of variation of the unemployment period is

$$C.V. = \sqrt{2(b/a) - 1}.$$

The probability density function of the unemployment period of an individual is

$$f(t) = (b-a)e^{-(b-a)t}, t > 0.$$

The probability that an individual is having an employment period less than or equal to $t$ is

$$U(t) = 1 - e^{-(b-a)t}.$$

The probability that an individual get a job at time $t$ is $b(t)$, which is complement of the unemployment period distribution function $G(t) = 1 - U(t) = e^{-(b-a)t}$.

The force of absorption can be measured by the rate of employment, which can be obtained as

$$Q(t) = -\frac{d}{dt}\log[G(t)] = f(t)/G(t) = f(t)/[1-f(t)]$$

$$= (b-a).$$

For this model, we observe that the force of absorption of the employment rate is constant, which is difference between the mean recruitment rate and the mean enrollment rate of the employment exchange.

## 4. The Joint Distribution of Number of Persons Recruited and Number of Persons Waiting for Job in an Employment Exchange

In this section we derive the distribution of the number of persons recruited through an employment exchange over a period of time (0, $t$). This is more important for developing proper recruitment policies in the organizations around the employment exchange. Here we assume that the enrollment process is Poisson with parameter *b*, and the recruitment is done through the employment exchange by first come first serve basis.

Let the recruitment process start with *i* persons enrolled for jobs in the exchange. Let the random variable *Y* (*t*) denote the number of persons already recruited through the employment exchange at time *t*. It is observed that *Y* (*t*) is monotonically increasing and represents the cumulative number of persons already got job through the exchange.

The joint probability, $P_{n,k}(t) = P[X(t) = n, Y(t) = k]$ ... (4.1)

The differential equations of the model are

$$\left.\begin{aligned} &P'_{i,o}(t) = -(a-b) - P_{i.o}(t) \\ &P'_{n,0}(t) = aP_{(n-1),0}(t) - (a+b)P_{n,0}(t), \text{ for } n > i. \end{aligned}\right\} \quad \ldots (4.2)$$

For $0 < k < i$, we obtain

$$\left.\begin{aligned} &P_{n,k}(t) = 0, && \text{for } n < (i-k) \\ &P'_{n,k}(t) + bP_{n+1,k-1}(t) - (a+b)P_{n,k}(t), && \text{for } n = (i-k) \\ &P'_{n,k}(t) = aP_{n-1,k}(t) - bP_{n+1,k-1}(t) - (a+b)P_{n,k}(t), && \text{for } n > (i-k) \end{aligned}\right\} \quad \ldots (4.3)$$

For $k \geq i$, we obtain

$$\left.\begin{aligned} &P'_{0,k}(t) = bP_{1.k-1}(t) - aP_{o.k}(t) \\ &P'_{n,k}(t) = aP_{n-1,k}(t) + bP_{n+1,k-1}(t) - (a+b)P_{n,k}(t) \end{aligned}\right\} \quad \ldots (4.4)$$

with the boundary conditions $P_{i,0}(0) = 1$ , $P_{n,0}(0) = 0$ for $(n, k) \neq (i, 0)$.

Consider the Laplace transformation of $P_{n,k}(t)$ as

$$P^*_{n,k}(s) = \int_0^\infty e^{-st} P_{n,k}(t)dt \, .$$

We have

$$P^*_{n,0}(s) = a^{n-i} / [(s+a+b)^{n-i+1}] \ (n \geq i, i > 0) \qquad \ldots (4.5)$$

We can invert (4.5) and obtain $P_{n,0}(t)$ for the set $0 < k < i$, from the generating function

$$A(x, y) = by \sum_{k=1}^{i-1} \sum_{n=i-k}^{\infty} P^*_{n,k}(s) x^n y^k \qquad \ldots (4.6)$$

Taking the Laplace Transform of (4.3) and using (4.6), we have

$$A(x, y) = \frac{\left[ by \sum_{n=i}^{\infty} P^*_{n,0}(s) x^n - by^i \sum_{n=i}^{\infty} P^*_{n,i-1}(s) x^n \right]}{\left[ x(s+a+b) - ax^2 - by \right]} \qquad \ldots (4.7)$$

From (4.5), we have $\sum_{n=i}^{\infty} P^*_{n,0}(s) x^n = x^i / (s+a+b-ax)$, which implies

$$A(x, y) = \frac{\left[ byx^i / (s+a+b-ax) - by^i \sum_{n=1}^{\infty} P^*_{n,i-1}(s) x^n \right]}{\left[ x(s+a+b) - ax^2 - by \right]} \qquad \ldots (4.8)$$

The zero in $x$ of the denominator in (4.8) is given by

$$x_0 (s+a+b) - ax_0^2 - by = 0 \qquad \ldots (4.9)$$

For $A\ (x, y)$ to remain finite for the value $x_0$, the numerator of (4.8) must have the same zero. Thus

$$\sum_{n=1}^{\infty} P^*_{n,i-1}(s) x_0^{\ n} = (x_0^{\ i} y^{i-1}) / (s+a+b-ax_0)$$

$$= (u^{i-1} x_0) / (s+a+b-ax_0)^i, \qquad \ldots (4.10)$$

where we have eliminated $y$ using (4.9). Because $x_0$ can be varied separately by changing $y$ we can expand the right side of (4.10) in power series of $x_0$ and, solve for $P^*_{n,i-1}(s)$. Thus we can replace $x_0$ by $x$ in (4.10), and (4.8) becomes

$$A(x,y)=\frac{[(byx^i)/(s+a+b-ax)-(b^i y^i x)/(s+a+b-ax)^i]}{\left[x(s+a+b)-ax^2-by\right]}. \quad \ldots (4.11)$$

This implies

$$A(x,y)=\left[\left(byx^{i-1}\right)/\left(s+a+b-ax\right)^2\right]\sum_{r=0}^{i-2}\left[\left(b^r y^r\right)/x^r\left(s+a+b-ax\right)^r\right].$$

Using the binomial expansion

$$A(x,y)=\frac{\sum_{k=1}^{i-1}\sum_{n=i-k}^{\infty}(2k+n-i)a^{n+k-i}b^k x^n y^k}{k!(n+k-i)!(s+a+b)^{2k+n-i+1}}.$$

Therefore

$$P^*_{n,k}(s)=\frac{\binom{n+2k-i}{k}a^{n+k-i}b^k}{(s+a+b)^{2k+n-i+1}},\ (k\le i-1,\ n\ge i-k). \quad \ldots (4.12)$$

Taking the Laplace Transform of (4.4), the generating function of $P_{n,k}(s)$ is

$$B(x,y)=\frac{\left[by^i\sum_{n=1}^{\infty}P^*_{n,i-1}(s)x^n-b(x-y)\sum_{k=1}^{\infty}P^*_{0,k}(s)y^k\right]}{x(s+a+b)-ax^2-by}. \quad \ldots (4.13)$$

Using (4.10) with $x_0$ replaced by $x$, we have

$$B(x,y)=\frac{\left[(b^i y^i x)/(s+a+b-ax)^i\right]-b(y-x)\sum_{k=1}^{\infty}P^*_{0,k}(s)y^k}{x(s+a+b)-ax^2-by} \quad \ldots (4.14)$$

when $x=x_0$, we have

$$(\mu^{\,i} y^i x_0)/(s+a+b-ax_0)^i = b(y-x_0)\sum_{k=1}^{\infty} P^{*}{}_{0,k}(s)y^k \quad \text{.... (4.15)}$$

Using (4.9),

$$(by)/(s+a+b-ax_0) = x_0 \quad \text{... (4.16)}$$

Therefore,

$$(x^{i+1})/[\mu(y-x_0)] = \sum_{k=1}^{\infty} P^{*}{}_{0,k}(s)y^k \quad \text{... (4.17)}$$

Thus (4.14) becomes

$$B(x,y) = \frac{[(b^i y^i x)/(s+a+b-ax)^i] - [(y-x)x^{i+1})/(y-x_0)]}{[x(s+a+b)-ax^2-by]} \quad \text{... (4.18)}$$

This implies

$$B(x,y) = \frac{[x^{i+1} - ((y-x)x^{i+1})/(y-x_0)]}{(x-x_0)(b-ax)} - \frac{[x^{i+1} - [(b^i y^i x)/(s+a+b-ax)^i]]}{[x(s+a+b)-ax^2-by]},$$

where

$$x_0 = [(s+a+b) - ((s+a+b)^2 - 4aby)^{1/2}]/2a$$

$$b = \frac{1}{2}[(s+a+b) - ((s+a+b)^2 - 4aby)^{1/2}]\,.$$

Dividing the first term of (4.18) by $(x-x_o)$ and expanding the second term and using (4.12), we have

$$B(x,y) = \left[\sum_{r=0}^{i-1} x_0^r x^{i-r} + (x_0^r y)/(y-x)\right]/(b-ax)$$

$$-\left[\sum_{r=o}^{i-1}\sum_{n=1}^{\infty}\binom{n+2k-i}{k} a^{n+k} b^k x^n y^k\right]/(s+a+b)^{n+2k-1+1} \quad \text{...(4.19)}$$

Taking $d = (by)/x_0$ and

$$c(x,y)=\left(\sum_{r=0}^{i-1} x_0{}^r x^{i-r}\right)/(d-ax).$$

This implies

$$c(x,y)=\left(\sum_{r=o}^{i-1}\ \sum_{n=i-r}^{\infty} a^{n+r-1} x^r x^{n+2r-i-1}\right)/\,(b^{n+r-i+1} y^{n+r-i+1}) \quad \ldots (4.20)$$

Using the Lagrange expansion theorem we have

$$x_0=\frac{\sum\limits_{t=0}^{\infty}\binom{2t+m-1}{t} a^t b^{t+m} y^{t+m}}{(s+a+b)^{2t+m}}-\frac{\sum\limits_{t=1}^{\infty}\binom{2t+m-1}{t-1} a^t b^{t+m} y^{t+m}}{(s+a+b)^{2t+m}} \quad \ldots (4.21)$$

After simplification, we have

$$c(x,y)=\frac{\left(\sum\limits_{n=1}^{i-1}\sum\limits_{k=i-n}^{\infty}\right)\binom{n+2k-i}{k} a^{n+k-i} x^n y^k b^k}{(s+a+b)^{n+2k-i+1}}$$

$$-\frac{\left(\sum\limits_{n=1}^{\infty}\sum\limits_{k=1}^{\infty}\right)\binom{n+2k-i}{k} a^{n+k-i} x^n y^k b^k}{(s+a+b)^{n+2k-i+1}}$$

$$+\frac{\left(\sum\limits_{n=1}^{\infty}\sum\limits_{k=0}^{\infty}\right)\binom{n+2k-i}{k} a^{n+k-i} x^n y^k b^k}{(s+a+b)^{n+2k-i+1}}. \quad \ldots (4.22)$$

Substituting (4.22) in (4.19) on simplification, we have

$$B(x,y)=\sum_{n=1}^{\infty}\sum_{k=i}^{\infty}\left[\binom{n+2k-i}{k}-\binom{n+2k-i}{k-i}\right]\left[\frac{a^{n+k-i} b^k x^n y^k}{(s+a+b)^{n+2k-i+1}}\right]$$

$$+x_0{}^i\, y/(y-x_0)(d-ax) \quad \ldots (4.23).$$

Let $D(x, y) = x_0^{\ i} y / (y - x_0)(d - ax)$.

$$D(x, y) = \frac{\left( \sum_{n=0}^{\infty} \sum_{j=0}^{\infty} \sum_{t=0}^{\infty} \binom{2t+n+j+i}{t} a^{n+t} x^n y^{t+i} b^{t+i+j} \right)}{(s+a+b)^{2t+n+j+i+1}}.$$

$$- \frac{\sum_{n=0}^{\infty} \sum_{j=0}^{\infty} \sum_{t=0}^{\infty} \binom{2t+n+j+i}{t-1} a^{n+t} x^n y^{t+i} b^{t+i+j}}{(s+a+b)^{2t+n+j+i+1}}$$

Taking $t+i=k$,

$$D(x, y) = \frac{\left( \sum_{n=0}^{\infty} \sum_{k=i}^{\infty} \sum_{j=0}^{\infty} \binom{2k+n+j-i}{k-i} a^{n+k-i} x^n y^k b^{k+j} \right)}{(s+a+b)^{2k+n+j-i+1}}$$

$$- \frac{1}{(s+a+b)^{2k+n+j-i+1}} \left( \sum_{n=0}^{\infty} \sum_{k=i+1}^{\infty} \sum_{j=0}^{\infty} \binom{2k+n+j-i}{k-i-1} a^{n+k-i} x^n y^k b^{k+j} \right) \quad \text{... (4.24)}$$

Taking the coefficient of $x^n y^n$, we find that

$$P_{n,k}(s) = \sum_{j=0}^{\infty} \left[ \binom{2k-i+j}{k-i} - \binom{2k-i+j}{k-i-1} \right]$$

$$\times \ a^{k-1} b^{k+j} / (s+a+b)^{2k-i+j+1}, \ k \geq i \quad \text{... (4.25)}$$

and

$$P_{n,k}^{\ *}(s) = a^{n+k-i} \ b^k \ / (s+a+b)^{n+2k-i+1}$$

$$\times \left[ \binom{n+2k-i}{k} + \sum_{j=1}^{\infty} \binom{n+2k-i+j}{k-i} b^j / (s+a+b)^j \right]$$

$$-\sum_{j=0}^{\infty}\binom{n+2k-i+j}{k-i-1} b^j/(s+a+b)^j \Bigg], n>0,\ k\geq i \qquad \ldots (4.26)$$

Taking the inverse transform of (4.5), we obtain

$$P_{n,0}(t)=(at)^{n-i}e^{-(a+b)t}/(n-i),\ n\geq i,\ i>0. \qquad \ldots (4.27)$$

From (4.12), we obtain

$$P_{n,k}(t)=[a^{n+k-i}b^k t^{n+2k-i}/k!(n+k-i)!]e^{-(a+b)t},$$

$$k\leq i\text{-}1,\ n\geq i\text{-}k, \ldots (4.28)$$

From (4.25), we have

$$P_{0,k}(t)=[(at)^{k-1}e^{-at}\gamma(k,bt)/(k-i)!(k-i)!]$$

$$-(a/b)(at)^{k-i-1}e^{-at}\gamma(k+1,bt)/(k-i-1)!k!],k\geq i., \ldots (4.29)$$

where $\gamma(\alpha, x)$ is the incomplete gamma function. We take

$$\gamma(\alpha, x)/(\alpha-1)! = e^{-x}\sum_{r=\alpha}^{\infty}(x^r/r!),\ \alpha=1,2,\ldots$$

$$=1,\ \alpha=0 \qquad \ldots (4.30)$$

Using $\gamma(\alpha+1,x)=\alpha\gamma(\alpha,x)-x^{\alpha}$, from (4.30) we obtain

$$P_{0,k}(t)=[a^{k-i}b^k t^{2k-i}/k!(k-1)!]e^{-(a+b)t}+[(at)^{k-i}/(k-i)!$$

$$-(a/b)(at)^{k-i-1}/(k-i-1)!]-e^{-at}[\gamma(k+1,bt)/k!],k\geq i.$$

From (4.28), we obtain

$$P_{n,k}(t)=[a^{n+k-i}b^k t^{n+2k-i}/k!(n+k-i)!]e^{-(a+b)t}$$

$$+[(at)^{k-i}/(k-i)!-(a/b)(at)^{k-i-1}/(k-i-1)!]$$

$$-(a/b)^n e^{-at}\gamma(n+k+1,bt)/(n+k)! \qquad \ldots (4.31)$$

$P_{n,k}(t)$ gives the joint probability that there are $n$ persons waiting for job in the employment exchange and $k$ persons have already being recruited through the employment exchange at time $t$.

## 5. The Distribution of Number of Persons Recruited during Time $t$

The marginal distributions $P_n(t)$ and $P_k(t)$ are obtained from summing the joint distribution $P_{n,k}(t)$. The result for $P_n(t)$ is well-known and it is the probability that there are $n$ in the system at time $t$. $P_k(t)$ is the probability that $k$ have already been retained during time $t$.

First for $k= 0$, we have, from (4.8), that

$$P_0^*(s) = \sum_{n=i}^{\infty} P_{n,0}^*(s) = 1/(s+b) \qquad \dots (5.1)$$

We have, from (4.9) and (4.14),

$$A(1,y) = \sum_{k=1}^{i-1}\sum_{n=i-k}^{\infty} P_{n,k}^*(s)y^k$$

$$= \left[by/(s+b) - b^i y^i/(s+b)^i\right]/(s+b-by).$$

Thus $A(1,y) = \sum_{k=1}^{i-1} b^k y^k /(s+b)^{k+1}$

and $P_k^*(s) = \sum_{n=i-k}^{\infty} P_{n,k}^*(s) = b^k/(s+b)^{k+1}, k \le i\text{-}1. \quad \dots (5.2)$

Taking inverse transform in (4.28), we have

$$P_k(t) = [(bt)^k/k!]e^{-bt}, k \le i\text{-}1. \qquad \dots (5.3)$$

Summing (4.31) for all values of $n$, we obtain

$$P_{0,k}(t) = [(bt)^k/k!]e^{-bt}[\gamma(k-1,at)/(k-i-1)!]$$

$$+[b/(b-a)][((at)^{k-i}/(k-i)!) - ((a/b)(at)^{k-i-1}/(k-i-1)!)]$$

$$\times[e^{-at}\{\gamma(k+1,bt)/k!\} - (b/a)^{k-bt}\{\gamma(k+1,at)/k!\}], k \ge i$$

$$\dots (5.4)$$

If $a = b$ in (4.31), we obtain

$$P_{0,k}(t) = [(at)^k e^{-at} / k!][\gamma(k-i, at)/(k-i-1)!]$$

$$+e^{-at}[(at)^{k-i}/(k-i)!-(at)^{k-i-1}/(k-i-1)!]$$

$$\times[(at)\gamma(k,at)-\gamma(k+1,at)/(k-1)!],\ k \geq i. \qquad \ldots (5.5)$$

The equation (5.5) gives the probability that $k$ have already been recruited at time $t$, through employment exchange. The equation (5.4) is also valid when $i= 0$, that is the inspection of the employment exchange. This probability distribution function is much useful for analyzing the performance and employment potentiality of the employment exchange, where it is situated.

## 6. Stochastic Model for Duration of Unemployment before Getting First Job

In this section we consider the stochastic modelling of unemployment duration associated with an employment exchange. Hence we assume that a person after completing the education, registers for a job in an employment exchange, which is having a specified recruitment rate.

Let $t$ be the unemployment duration of the person from his enrollment in the exchange to get his first job. For developing the suitable model consider the following postulates :

(1) There exists a function $m$, such that for any interval $(t_1, t_2)$ with $0 < t_1 < t_2$ and for $n= 0,1,2\ldots,$ the probability that an enrolled person has $n$ trials during the time interval $(t_1, t_2)$ for an employment is Poisson. Then

$$P_n(t_1, t_2) = [A(t_1, t_2)]^n e^{-A(t_1,t_2)} / n!,$$

where $A(t_1, t_2) = \int_{t_1}^{t_2} m_1 dt$.

(2) The events in non-overlapping intervals of time are statistically independent.

(3) The probability that a trial results in getting job is constant and denote it by $P_1$. It can be seen that the number of such recruitments lying in the interval $(t_1, t_2)$ follows a Poisson distribution with parameter $m' = m_1 p_1$.

(4) The probability that a recruitment leads to a permanent job is *b*. Thus the number of permanent jobs follows a Poisson distribution with parameter $k = m'(b) = m_1 p_1(b)$.

(5) If there is permanent employment, then the probability of another trial for other job in a constant time *h* is zero.

(6) All persons under consideration are in the age group of work force.

Let *O* be the random variable denoting the duration of unemployment period. Let us assume that the unemployment under study has *c* recruitments during the period (*O*, *T*). The $i^{th}$ recruitment is occurred at $(T-t)$ from the enrollment of a person. So, *t* is the length of *O*, i.e., the duration of unemployment for a person at time *T* which is the time of observation.

Let *i* successive trials occurs for getting a job by a person at times $x_1, x_2, x_3, \ldots$, respectively. Then $x = (T-t-h)$, since *h* is constant for all trials $x_i - 1$ lies anywhere in the interval [ $(i-2)/h$, $(T-t-h)$], $x_i-2$ lies anywhere in the interval [ $(i-3)$

$/h, (x_i - 1 - h)]$, $x_i$ lies anywhere in $(0, x_2 - h)$, since $i$ trials occurs successfully at times $x_1, x_2, \ldots, x_i$, no trial occurs during the period $(0, x_1), (x_1 + h, x_2), \ldots, (x_{i-1} + h, x_i)$.

Thus the probability that the least trial takes the period during the interval $(x_1, x_1 + dx_1)$ is $e^{-t(0,x_1)} k dx_1$. Therefore, the probability that there is no trial in the period $(x_1 + h, x_2)$ and the second trial occurs during $(x_2, x_2 + dx_2)$.

Similarly, the probability that there is no trial during the period $(x_j + h, x_{j+1})$ $j = 2, 3, 4, \ldots, (i-1)$ and $(j+1)^{th}$ trial occurs in a small interval $(x_{j+1}, x_{j+1} + dx_{j+1})$ is $e^{-(x_j + h, x_{j+1})} k dx_{j+1}$.

Since all trials are independent, the joint probability that $i$ successive trials occur at times $x_1, x_2, \ldots, x_i$, from the date of enrollment is

$$\int\limits_{x_{i-1}} \cdots \int\limits_{x_2} \int\limits_{x_1} e^{-t(0,x_1) - t(x_1 + h, x_2) - \ldots - t(x_{i-1} + h, x_i)} k^i d_{x_1} d_{x_2} \ldots d_{x_i} .$$

Since $x_i = T - t$, $dx_i = -dt$, where $dt$ is negative under the assumption (4) is

$$A(t_1, t_2) = \int_{t_1}^{t_2} k dt \qquad \ldots (6.1)$$

Therefore, by integrating the above integral over $x_1, x_2, \ldots, x_{i-1}$ the probability that $i$ successive recruitments occur, the $i^{th}$ recruitment occurring in the interval $(T-t, T-t+dt)$, $dt$ being positive, is

$$\frac{k^i [(T - t - (i-1)h]^{i-1}}{(i-1)!} e^{-k[T-t-(i-1)h]} dt .$$

Combining with the probability that no recruitment occurs in the interval $[T-t-(i-1)h]$, the probability that zero lies between $t$ and $t+dt$ is

$$\frac{k^i[(T-t-(i-1)h]^{i-1}}{(i-1)!}e^{-k[T-t-(i-1)h]}dt = f_2(t)dt \quad \ldots (6.2)$$

and the probability that zero lies between $t$ and $t+dt$ for $h \le t \le (T-(i-1))h$ is

$$\frac{k^i[(T-t-(i-1)h]^{i-1}}{(i-1)!}e^{-k[T-ih]}dt = f_1(t)dt\,. \quad \ldots (6.3)$$

Hence the probability density function of the duration of unemployment after the $i^{th}$ recruitment is obtained as

$$f(t) = \frac{k^i[(T-t-(i-1)h]^{i-1}}{(i-1)!}e^{-k[T-ih]}, h \le t \le (T-(i-1))h$$

$$= k^i[(T-t-(i-1)h]^{i-1}e^{-k[T-t-(i-1)h]}, 0 \le t \le h \ldots \quad (6.4)$$

Using the probability density function, we can obtain the average duration of the unemployment period after the $i^{th}$recruitment as

$$E(t) = \int_0^h tf(t)dt + \int_h^{t-(i-1)h} tf(t)dt$$

$$= [(i/k) - T - (i-1)h]F[k(T-ih)] - F[k(T-(i-1)h)]$$
$$+ [k^i/(i+1)!]e^{-k(T-ih)}(T-ih)^i[T-(2i+1)h]\cdot$$

$$\ldots (6.5)$$

The variance of the unemployment duration is obtained as

$$V(t) = E(t^2) - [E(t)]^2$$

$$
\begin{aligned}
&= -\{F(k(T-ih)F[k(T-(i-1)h]\}[(T_i/(i-1)!] \\
&\{(t-(i-1)h)^2 + [i(i+1)/k^2] - 2i[(F(i-1)h)/k] \\
&+[(i/k)-(T-(i-1)h)]^2[F(k(T-ih)-F(k(T-(i-1)h)] \\
&+(k^i/i!)e^{-k(T-ih)}(T-ih)^i\{h^2+(2h(T-ih)/i(i+1) \\
&+(2(T-ih)^2/i(i+1)(i+2))\}+(k^i/(i+1)!)e^{-k(T-ih)}(T-ih)^i(T-(2i+1)h) \\
&\times[(k^i/(i+1)!)e^{-k(T-ih)}(T-ih)(T-(2i+1)h) \\
&+2((1/k)-(T-(i-1)h)(F(k(T-i-h)))-F(k((T-(i-1)h))]
\end{aligned}
\quad \ldots (6.6)
$$

By using this model one can analyze the unemployment duration associated with his first permanent job. This model is also useful to analyze the number of interviews to be attempted before getting a permanent job after enrolled in the employment exchange.

## 7. Conclusion

This paper deals with the stochastic modelling associated with an employment exchange. Here we consider that the unemployed individuals enrolled at a specific employment exchange, for jobs in Poisson fashion. The recruitment of individuals in that region is through the employment exchange. Assuming that the recruitment process is Markovian, the number of employed individuals in that region is modelled through the differential equations. Assuming the system is in steady state the growth of unemployment in the region is analyzed by deriving the distributions of number of unemployed individuals remains in the employment register and the waiting time for getting recruitment. It is observed that the waiting process in the employment exchange is of gamma type. The characteristics of the model like, the average number of unemployed individuals in the region at any time, the average waiting time of an individual after being enrolled in the employment exchange, the variability of the unemployment duration, the coefficient of variation of the unemployment duration, the force of absorption into jobs, the rate of employment are derived and analyzed. It is observed that the force of absorption or employment rate of the specific region is the difference between the mean recruitment rate and mean enrollment rate of the employment exchange. For appropriate analysis of this problem, it is desirable to have the analysis as time dependent, so the transient solution of the above model is also derived by assuming that there are "*i*" persons already registered for job in the employment exchange at time $t = 0$. The distribution of number of persons recruited through the employment exchange for a period of time is also derived by using the joint probability concepts. This distribution is very useful for developing suitable employment programmes in the region. Incidentally, it is also observed that the enrollment rate of an employment exchange is heavily influenced by the recruitment rate of the region. This deduction is formed through the joint probability distribution of "*n*" persons waiting for the job in the employment exchange and " *k*" persons have already recruited through the employment exchange at time "*t*". Another stochastic model is developed for the duration of unemployment before getting his first job. Here we assume that a person after completing education registers for a job in an employment exchange, which is having a specified recruitment rate. We further assume that the probability that the enrolled person has attempted "*n*" trials during the time interval $(t_1, t_2)$ for an employment is Poisson and the probability density function of the duration of unemployment after $i^{th}$ recruitment is derived. These models are useful for assessing the economic development of the regions.

## References

Aitkin, M. and Healey, R. (1985), Statistical Modeling of Unemployment Rates from the EEC Labour Force Survey, *JRSS, Series A*, 148.

Baily, M.N. (1977), On the Theory of Layoffs and Unemployment, *Econometrica*, 45.

Cargill, T. F. and Meyer, R.A. (1974), Wages, Prices and Unemployment : Distributed Lag Estimates, *JASA*, 69.

Casler, V. (1988), The Estimation of the Distribution of Unemployment Spells in Spain.

Caster, T.L. (1979), Econometric Methods for the Duration of Unemployment, *Econometrica*, 47 (4).

Caster, T.L. and Nickell, S. (1980), The Analysis of Re-employment Probabilities for the Unemployed, *JRSS, Series A*, 143.

Frank, R.H. (1978), How Long is a Spell of Unemployment, *Econometrica*, 46.

Horowitz (1970), Employment Concentration in the Common Market : An Entropy Approach, *JASA*, 65.

James, J.A. (1995), Reconstructing the Pattern of American Unemployment before the First World War, *Econometrica*, 62.

Korpi, T. (1995), Effects of Manpower Policies on Duration Dependence in Re-employment Rates, *Econometrica*, 62.

Maria, E.G. and Christinehoza (1978), Small-Area Estimation with Application to Unemployment and Housing Estimates, *JASA*,73.

Meyer, B.D. (1990), Unemployment Insurance and Unemployment Spells, *Econometrica*, 58 (4).

Moffit (1985), Unemployment Insurance and Distribution of Unemployment Spells, *Journal of Econometrics*, 28.

Morgenstern, R.D. and Barrett, N.S. (1974), The Petrospective Bias in Unemployment Reporting by Sex, Race and Age, *JASA*, 69.

Oliver (1964), Inter-regional Migration and Unemployment, 1951-1961, *JRSS, Series A*, 127.

Storer, P. (1994), Unemployment Dynamics and Labour Market Evaluation of Matching Function Models, *Journal of Applied Econometrics*, 9.

Uzuwa (1964), Involuntary Unemployment and Growth Equilibrium, *Econometrica*, 32 (3).

9

# ARCH Processes and Their Applications in Stock Return Data

***A.K. Basu*** **and** ***Sankha Bhattacharya***

## 1. Introduction

Ever since its early days, an important application of time series models has been in the field of finance. This has ranged from the studies on growth parameters of an economy to the stock markets and portfolio management. However, such series have their own peculiarities and hence require modifications of the usual time series models. In this article, we seek to look at some of the developments in time series analysis that have been induced by financial models.

A financial time series features various forms of non-linear dynamics, the most crucial one being the strong dependence of the instantaneous variability of the series on its own past. The traditional econometric models used to assume a constant one period forecast variance. However, it was recognized for quite some time that the uncertainty of speculative prices, as measured by the variances and co-variances, change through time [Fama, 1965]. Moreover, financial theories are based on concepts like equilibrium or rational behaviour of the investors, which naturally suggests including and testing some structural constraints on the parameters. Thus the so-called ARMA process failed to fit in such problems because of its linear set-up and unconstrained AR and MA parameters. In this context the Autoregressive

Conditional Heteroscedastic (ARCH) model introduced by Engle (1982) serves as one of the most prominent tools to characterize such changing variances.

## 2. Autoregressive Conditional Heteroscedastic (ARCH) Model

Engle (1982) assumed that the conditional forecast variance $V(Y_t/Y_{t-1})$ depends upon the past thereby assuming that it itself is a random variable. His proposed model was

$$Y_t = \varepsilon_t h_t^{1/2} \quad \text{with } V(\varepsilon_t) = 1, \qquad \ldots (2.1)$$

$$h_t = h(y_{t-1}, y_{t-2}, \ldots, y_{t-p}, \alpha), \qquad \ldots (2.2)$$

where

$$Y_t \mid \psi_{t-1} \sim N(0, h_t) \qquad \ldots (2.3)$$

$\psi_t$ is the information set available at time $t$, $p$ is the order of the ARCH process and $\alpha$ is a vector of unknown parameters. One possible parametrization for $h_t$ suggested by Engle (1982) is to express $h_t$ as a linear function of past squared values of the process, i.e.,

$$h_t = \alpha_0 + \sum_{i=1}^{p} \alpha_i y_{t-i}^2,$$

where $\alpha_0 > 0$, $\alpha_i \geq 0$.

This model is known as linear ARCH($p$) model. He also proposed arch regression model in his paper by assuming that the mean of $y_t$ is given by $x_t\beta$, a linear combination of lagged endogenous and exogenous variables included in the information set $\psi_{t-1}$ with $\beta$, a vector of unknown parameters. Formally

$$Y_t \mid \psi_{t-1} \sim N(x_t\beta, h_t)$$

$$h_t = h(\varepsilon_{t-1}, \varepsilon_{t-2}, \ldots \varepsilon_{t-p}; \alpha),$$
$$\varepsilon_t = Y_t - x_t\beta.$$

If $h_t$ is a $p^{th}$ order linear in the squares it can be written as

$$h_t = \alpha_0 + \alpha_1 \varepsilon^2_{t-1} + \ldots + \alpha_p \varepsilon^2_{t-p}.$$

The log-likelihood of the $t^{th}$ observation is given by

$$l_t = -1/2 \log h_t - ½ \ \varepsilon^2_t / h_t.$$

The log-likelihood can be maximized with respect to the unknown parameters $\alpha$ and $\beta$. The maximum likelihood based inference procedure for ARCH class of models under the assumption of normal distribution were discussed in Engle (1982) and Pantula (1985). Although the likelihood function is highly non-linear in parameters, a simple scoring algorithm is available for the linear ARCH($p$) model defined in (2.1) - (2.3). Each step for a parameter vector $\phi$ produces estimates $\phi^{(i+1)}$based on $\phi^{(i)}$according to

$$\phi^{(i+1)} = \phi^{(i)} + [\vartheta^{(i)}_{\phi\phi}]^{-1}.(1/T). \Sigma_i \delta l_t^{(i)} / \delta\phi,$$

where $\vartheta^{(i)}$ and $l_t^{(i)}$ are evaluated at $\phi^{(i)}$. $\vartheta^{(i)}_{\phi\phi}$ is the Hessian averaged over all observations for estimating $\phi$. Engle (1982) used this model to estimate the means and variances of inflation in UK. The ARCH effect was found to be significant and the estimated variances increased significantly during the chaotic seventies.

Further, in Engle (1982, 1983), models for the inflation rate were constructed recognizing that the uncertainty of inflation tends to change over time. Weiss (1984) used ARMA models with ARCH errors to successfully model thirteen different U.S. macroeconomic time series. He also showed that ignoring ARCH would lead to the identification of

ARMA models that are over-parametrized. Engle (1982) presented the Lagrange Multiplier (LM) test to identify the presence of an ARCH model. That is a test of the null hypothesis $H_0: \alpha_1 = \alpha_2 = \ldots = \alpha_p = 0$. The test procedure is based on the test statistic $T.\, R^2$, which is a $\chi^2_p$-variate, where $R^2$ is the squared multiple correlation coefficient of the regression obtained from the regression of $\varepsilon^2_t$ on $\varepsilon^2_{t-1}, \varepsilon^2_{t-2}, \ldots \varepsilon^2_{t-p}$ and $T$ is the sample size.

## 3. Generalized Auto-regressive Conditional Hetero-scedastic (GARCH) Process

In 1986, Bollerslev proposed a more general class of processes known as Generalized Auto-regressive Conditional Heteroscedastic (GARCH) process, which allows a much more flexible lag structure. Engle (1982) explicitly recognized the difference between the unconditional and conditional variances keeping the former constant over time and the latter to change over time as a function of past errors. In the ARCH ($p$) process the conditional variance is specified as a linear function of past sample variances only, whereas the GARCH ($p$, $q$) process allows lagged conditional variances to enter as well.

The GARCH ($p$, $q$) model is thus given by

$$\varepsilon_t \mid \Psi_{t-1} \sim N(0, h_t) \qquad \ldots (3.1)$$

$$h_t = \alpha_0 + \sum_{i=1}^{p} \alpha_i \varepsilon^2_{t-i} + \sum_{i=1}^{q} \beta_i h_{t-i}, \qquad \ldots (3.2)$$

where $p \geq 0, q > 0, \alpha_0 > 0, \alpha_i \geq 0$, for $i = 1, \ldots, p$, $\beta_i > 0$ for $i = 1, \ldots, q$.

The GARCH ($p,q$) regression model is obtained by using $\varepsilon_t$'s as innovations in a linear regression model,

$$\varepsilon_t = Y_t - x_t b,$$

where $y_t$ is the dependent variable, $x_t$ a vector of explanatory variables, $b$ is a vector of unknown parameters. The log likelihood function for a sample of $T$ observations (excluding terms which are constant) is

$$L_t(\theta) = T^{-1} . \sum_{t=1}^{T} l_t(\theta)$$

$$l_t = -1/2 \log h_t - ½\, \varepsilon^2_t / h_t$$

Owing to the presence of the recursive term in the expressions of $\delta h_t / \delta\omega$ and $\delta h_t / \delta b$, where $\omega=(\alpha_0, \alpha_1, \ldots, \alpha_p, \beta_1, \beta_2, \ldots \beta_q)'$ and $b$ as defined above, the application of the method of scoring is not, unlike in the ARCH($p$) model, suitable for this case. Let $\theta^{(i)}$ denote the parameter estimates after the $i^{th}$ iteration. $\theta^{(i+1)}$ is then calculated from

$$\theta^{(i+1)} = \theta^{(i)} + \lambda_i . [\sum_{t=1}^{T} (d\,l_t / d\theta)(d\,l_t / d\bar{\theta})]^{-1} . \sum_{t=1}^{T} d\,l_t / d\theta,$$

where $dl_t / d\theta$ is evaluated at $\theta^{(i)}$ and $\lambda$ is a variable step length chosen to maximize the likelihood function in the given direction. What can be tested is the null hypothesis of an ARCH($p$) process against a GARCH($p, q$) alternative, i.e., $H_0$: $\beta_1=\beta_2=\ldots\beta_q=0$. Bollerslev (1986,1987) showed that the appropriate test statistic is $T.\ R^2$ computed from the regression of $(\varepsilon^2_t h_t^{-1}-1)$ on the vector $(h_t \delta h_t / \delta\omega)$, both being evaluated under $H_0$, i.e., the regressors are $(h_t z_t)$, where $z_t$ is given as $z'_t = (1, \varepsilon^2_{t-1}, \ldots, \varepsilon^2_{t-p}, h_{t-1}, \ldots, h_{t-q})$.

## 4. The GARCH (1,1) Process

The simplest but often very useful GARCH process is the GARCH (1,1) process given by

$$h_t = \alpha_0 + \alpha_1 \varepsilon^2_{t-1} + \beta_1 h_{t-1}, \qquad \ldots (4.1)$$

$\alpha_0 > 0, \alpha_1 \geq 0, \beta_1 \geq 0, \alpha_1 + \beta_1 < 1.$

The mean lag in the conditional variance equation is given by $(1-\beta_1)^{-1}$ and the median lag by-$\log_1^2/\log\beta_1$. The coefficient of kurtosis is greater than one and hence GARCH (1,1) process is leptokurtic.

We have mentioned above that the stationarity of GARCH (1,1) process is ensured if $\alpha_1 + \beta_1 < 1$.

For $s \geq 2$,

$$E(h_{t+s} | \varepsilon_t, \varepsilon_{t-1}, ...) = \alpha_0 + (\alpha_1 + \beta_1) E(h_{t+s-1} | \varepsilon_t, \varepsilon_{t-1}, ...) \quad ... (4.2)$$

Suppose now $\alpha_0 = 0$ and $(\alpha_1 + \beta_1) = 1$,

Then

$$h_t = \alpha_1 \varepsilon^2_{t-1} + (1-\alpha_1) h_{t-1} \quad .... (4.3)$$

$$E(h_{t+s} | \varepsilon_t, \varepsilon_{t-1}, ...) = E(h_{t+s-1} | \varepsilon_t, \varepsilon_{t-1}, ...) = E(h_{t+1} | \varepsilon_t, \varepsilon_{t-1}, ...) = h_{t+1} \quad ... (4.4)$$

The conditional variance of *s*-step ahead predictor is same as the conditional variance of one step ahead predictor for all horizons *s*, and the model is obviously closely related to the traditional random walk, which has a unit root in the conditional mean rather than as here, a unit root in the conditional variance. This type of model is known as integrated GARCH, IGARCH (1,1) model (Engle and Bollerslev, 1986). The implication of the model is that effect of $h_{t+1}$ on the future conditional variances is permanent.

In the basic ARCH model the distribution of $\varepsilon_t$ was assumed to be conditionally normal. Bollerslev (1987) considers the case when the distribution is standardized-*t* with unknown degrees of freedom that can be estimated from the data : for $\nu > 4$ such a distribution is leptokurtic and hence has 'fatter tails' than the normal distribution. Other distributions that have been considered include the normal-Poisson mixture distribution in Jorian (1988), the power-exponential distribution in Baillie and Bollerslev (1989), the normal-lognormal mixture in Hsieh (1989), and the generalized exponential distribution in Nelson (1991).

Engle and Bollerslev (1986) considered two variants to the GARCH (1,1) model

$$h_t = \alpha_0 + \beta_1 h_{t-1} + \alpha_1 |\varepsilon_{t-1}| \lambda^1 \quad \text{... (4.5)}$$

and

$$h_t = \alpha_0 + \beta_1 h_{t-1} + \alpha_1 ( 2\Phi(\varepsilon^2_{t-1}/\lambda_2) - 1 ), \quad \text{... (4.6)}$$

where $\Phi$ denotes the standard normal cumulative distribution function. If $\lambda_1 = 2$ in (4.5), then the model is just GARCH(1,1) while if in (4.6), $\lambda_2$ is large relative to $\sigma$, the model is essentially linear. Geweke (1986) and Pantula (1986) suggested a logarithmic parametrization, while Higgins and Bera (1992) generalized further by considering a specification in which both $h_t$ and $\varepsilon_t$ s are transformed using the Box and Cox power transformation.

## 5. Exponential GARCH (EGARCH)

Nelson (1991) and Pagan and Schwert (1990) considered another variant, the exponential GARCH (EGARCH) model, in which $h_t$ is an asymmetric function of past $\varepsilon_t$ s. Formally EGARCH model is given by

$$\ln(h_t) = \alpha_0 + \beta_1 (h_{t-1}) + \alpha_1[\theta U_{t-1} + (| U_{t-1}| - (2/\pi)^{1/2}],$$

where $U_t = \varepsilon_t / h_t$.

## 6. ARCH in MEAN Model (ARCH-M Model)

Engle et al. (1987) introduced ARCH in Mean model, which extends the ARCH model to allow the conditional variance to affect the conditional mean. In this way the changing conditional variance directly affect the expected return on a portfolio. In such models

$$Y_t \mid \Psi_{t-1} \sim N(\beta x_t + \partial h_t, h_t)$$

$$h_t = \alpha_0 + \sum_{i=1}^{p} \alpha_i \varepsilon_{t-i}^2,$$

where $E(Y_t \mid \Psi_{t-1})$ gives the conditional mean of $Y_t$ given the past observations and $h_t$ gives the conditional variance function of the information available to the investor. This model has been used further to estimate the approximate linear relationship between the return and variance of market portfolio. They used ARCH-M model where the conditional variance is a determinant of the current risk premium, and thus enters the forecasting equation of expected financial returns.

## 7. Application of ARCH to Stock-Return Data

Many studies have shown that time series of daily stock returns exhibit some auto-correlation for short lags. The magnitudes of the auto-correlation are too small to form any profitable trading rule. Moreover, the empirical distributions of the returns are significantly different from the Gaussian distribution. As a result the usual auto-correlation estimates may be even low. Again, unless the underlying stochastic process is Gaussian, the lack of serial correlation does not imply statistical independence. The discussions and empirical evidences given by many authors like Pindyck (1984), Poterba and Summers (1986) further imply that the assumption of constant conditional means and variances is both unrealistic and dubious. So all that is required is to express the return generating process as martingale.

In 1974, Black found that stock returns are negatively

correlated with changes in volatility. Malkeil (1979) and Pindyck (1984) argued that upward trend in volatility during the past three decades is the major reason for the decline in equity prices. Pindyck agreed that the "variance of the firms' real gross marginal return on capital has increased the relative risk of investors' net real returns from holding stocks, and that this, in turn, can explain a large fraction of market decline."

Poterba and Summers (1986), on the course of examining Black's finding by testing the time series property of volatility, argued that shocks to volatility have to persist for a very long time in order for volatility to have significant effect on stock prices. If these shocks are only transitory, no adjustment of the future discount rate is made by the market. Thus expected stock returns are not affected by the volatility movement. Hence by empirically rejecting the persistence property using some volatility measures, they refuted the above claim of Malkeil (1979) and Pindyck (1984). They concluded that shocks to the stock volatility are very short-lived. As a result, using the fundamental valuation formula, they obtained a very small estimate for the elasticity of the stock price to volatility shocks.

French et al. (1987) derived a series of anticipated volatility from the fitted values of an ARIMA process which was found to be weakly related to the expected risk premium. On the other hand, strong evidence was found that 'unexpected stock market returns are negatively related to the unexpected volatility of stock returns'.

Pindyck (1986) rejoined the issue by estimating a portfolio choice model. He obtained estimates of the index of relative risk aversion in the range of 3 to 4. While admitting that changes in variance do not persist long, he still claimed that 'they do seem to explain more than other variables (e.g., changes in corporate profits and changes in the real interest rate), because they have been large in magnitude and because the index of risk aversion is large'.

Chou (1988) used a univariate GARCH-M model to study the issue of volatility persistence and its relationship with market fluctuations. He employed the fundamental evaluation formula following Poterba and Summers to evaluate the impact of volatility on the stock prices. Like Pindyck (1986), he also estimated the index of relative risk aversion in a maximum likelihood estimation framework. The expected rate of return from time $t$ to $t$+1 of the market portfolio with dividend reinvested denoted by $R_{t+1}$, is the sum of two components, the risk free rate and a risk premium.

$$E_t(R_t) = [E_t(S_{t+1}) - S_t + D_t]/ S_t = r_{f,t} + p_t, \quad \dots (7.1)$$

where $S_t$ , $D_t$ , $r_{f,t}$ , $p_t$ are, respectively, the stock price, dividend paid, nominal risk-free rate and risk premium at period $t$.

Marton (1973, 1980) gave an inter-temporal CAPM model which implies a linear relationship between the equity premium and market return variance, i.e., $p_t = \delta V_t$, where $V_t$ is the instantaneous variance of the market return and $\delta$ is the harmonic mean of individual investors' Pratt-Arrow measure of relative risk aversion. An AR(1) specification has been adopted by Poterba and Summers (1986) and Pindyck (1984) as the specification of the volatility series $V_t$. A two-stage method was used for estimation and this gave a very unreliable result.

Chou (1988) used the GARCH(1,1) specification for the volatility series given by

$$V_t = \alpha_0 + \alpha\varepsilon^2{}_t + \beta V_{t-1}, \quad \dots (7.2)$$

where $\varepsilon_t$ is the forecasting error for the stock return,

$$\varepsilon_t = R_t - E_{t-1}(R_t),$$

where $V_t = \text{Var}(R_{t+1}/I_t)$ is the conditional variance of return $R_{t+1}$ given information up to time $t$. From equation (7.2),

$$E_t(V_{t+s}-\sigma^2) = (\alpha+\beta)^s(V_t-\sigma^2),$$

where $\sigma^2$ is the unconditional variance of $R_t$ given by

$$\sigma^2 = \alpha_0/[1-(\alpha+\beta)] \text{ for } \alpha+\beta<1.$$

The reason for which Chou (1988) chose GARCH(1,1) specification is because GARCH(1,1) has the convenient property that shocks to volatility decay at a constant rate and the speed of decay is measured by the estimate of $\alpha+\beta$ (specification parameter, assumed less than 1 for his case). He proceeded to estimate the model parameters by defining a GARCH (1,1)-M model on the basis of conditions (7.1), (7.2), and (7.3),

$$R_{t+1} = r_f + \delta.V_t + \varepsilon_{t+1}$$

$$V_t = \alpha_0 + \alpha\varepsilon^2_t + \beta V_{t-1}$$

$$\varepsilon_{t+1} | I_t \sim N(0,V_t).$$

He used the model to fit the data of the weekly return of the NYSE value weighted index with dividends reinvested. The sample period was from July 1962 to December 1985 and a total of 1225 observations were available. He also made a comparison with AR(1) two-stage estimates and concluded that GARCH-M estimation method provides a more stable estimate than the AR(1) two-stage method of Pindyck.

Application of the GARCH-M model was carried out by other authors which includes French et al. (1987) for the daily *S* and *P* index, Friedman and Kuttner (1988) for quarterly U.S. stock indices. Some evidences of sensitivity of the parameter estimates in the ARCH-M model with respect to different model specifications is given by Baillie and De Gennaro (1990) who used both daily and monthly portfolio returns. To analyze the volatility of daily U.S. stock returns they used the model

$$Y_t = \alpha + \beta x_t + \delta h_t^{\lambda} + U_t,$$

where

$$U_t = \varepsilon_t - \theta.\varepsilon_t$$
$$h_t = \nu_0 + \nu_1 \varepsilon^2_{t-1} + \phi h_{t-1} + \vartheta \xi_t,$$

$\varepsilon_t \mid \Psi_{t-1}$ follows *t* distribution with ν degrees of freedom with mean 0 and variance $h_t$.

In this model $Y_t$ is the actual return on day *t*, $x_t$ is a variable measuring the compensation received by the sellers of stock for payment delays caused by settlement and cheque-clearing procedures, which being proxied by the change from day *t-1* to *t* in the federal funds rate continuously compounded over the delay period, and $\xi_t$ is the absolute value of this change, i.e., $\xi_t = |x_t|$. The inclusion of $h_t^{\lambda}$ is an attempt to incorporate a measure of risk into the return generating process.

With risk averse investors α is expected to be positive as is β, since compensation should increase as the payment delay period increases. Under the mean variance hypothesis $\delta > 0$ so that large values for the conditional variance are expected to be associated with large returns.

Baillie and DeGennaro (1990) estimated this model using daily returns data from 1st January, 1970 to 22 December, 1987, a total of 4542 observations. They concluded that any relationship between mean returns and return variance is weak and that the results suggested that investors consider some other risk measure to be more important than the variance and portfolio returns. Similar results are found in French et al. (1987), Bollerslev and Wooldridge (1991) using robust standard error. Gallant et al. (1990) found similar results in a semi-parametric framework.

## 8. Conclusion

In general, the family of GARCH models has proved to be a popular method of modelling volatility in financial markets.

Nelson (1991) however discusseed some limitations that GARCH models have in this context. One such limitation follows from the so-called leverage effect (Black (1976)). According to the leverage effect the changes in stock return volatility is negatively correlated with returns themselves, i.e., the volatility tends to rise in response to 'bad news' (returns lower than expected) and to fall in response to 'good news'(return higher than expected). The general GARCH $(p, q)$ model is not able to capture this kind of dynamic pattern since the conditional variance $h_t$ is only linked to past conditional variances and squared innovations ($\varepsilon^2_t$' s) that affect $h_t$ in equation (3.2) so that $h_t$ is invariant to the sign of $\varepsilon_t$'s.

A second important limitation of GARCH models is the non-negativity constraints placed on the parameters of the GARCH process to ensure that $h_t$ remains non-negative for all $t$. This constraint implies that increasing $\varepsilon^2_t$'s in any period increases $h_{t+m}$ for all $m \geq 1$, thus ruling out random oscillatory behaviour in the $h_t$ process. Furthermore, these non-negativity constraints can create problems when estimating GARCH models and it may be difficult to determine whether shocks in GARCH models are persistent or not.

Nelson (1991) presented an alternative to the GARCH model that meets these objectives. To ensure that $h_t$ remains non-negative, but without imposing inequality constraint on model parameters, Nelson assumed that natural logarithm of $h_t$ is a function of lagged $U_t$ where $U_t = \varepsilon_t / h_t$.

His proposed model (known as EGARCH) is

$$Y_t = \alpha_0 + \alpha_1 Y_{t-1} + \delta h_t + \varepsilon_t$$

$$\ln(h_t) = v_0 + \phi_1 U_{t-1} + \phi_2 [\,|U_{t-1}| - E|U_{t-1}|\,]$$
$$+ \phi_1 \ln(h_{t-1}) + \phi_2 \ln(h_{t-2}) + \vartheta \xi_t.$$

The model was fitted to daily US stock returns from July

1962 to December 1987. Here $\xi_t$ is a variable measuring the influence of non-trading periods on the conditional variance and $U_t$ is assumed to be generated as i.i.d. observations, drawn from a generalized error distribution. The estimated risk premium $\delta$ is negative but insignificant. The estimates of $\phi_1$ and $\phi_2$ are significantly negative and positive, respectively, and this provides an asymmetric relationship between returns and changes in volatility. All the major episodes of high volatility are associated with market drops and the estimates of the auto-regressive parameters $\phi_1$ and $\phi_2$ imply a largest root of almost unity, indicating substantial persistence of shocks to the conditional variance.

## References

Baillie, R.T. and Bollerslev, T. (1989), The Message in Daily Exchange Rates : a Conditional Variance Tale, *Journal of Business and Economic Statistics*, 7, 297-305.

Baillie, R.T. and DeGennaro, R. P. (1990), Stock Returns and Volatility, *Journal of Financial and Quantitative Analysis*, 25, 203-214.

Black, F. (1976), Studies of Stock Price Volatility Changes, *Proceedings of the 1976 Meeting of Business and Economic Statistics Section*, American Statistical Association, 177-181.

Bollerslev, T. (1986), Generalized Autoregressive Conditional Heteroscedasticity, *Journal of Econometrics*, 31, 307-327.

Bollerslev, T. (1987), A Conditionally Heteroscedastic Time Series Model for Speculative Prices and Rates of Return, *Review of Economics and Statistics*, 69, 542-546.

Bollerslev, T. (1987), On Correlation Structure for the Generalized Autoregressive Conditional Heteroscedastic Process, *Journal of Time Series Analysis*, 9, 121-132.

Bollerslev, T. and Wooldridge, J. M. (1991), Quasi Maximum Likelihood Estimation and Inference in Dynamic Models with Time Varying Co-variances, *Econometric Reviews*.

Chou, R.Y. (1988), Volatility Persistence and Stock Valuation : Some Empirical Evidence Using GARCH, *Journal of Applied Econometrics*, 3, 279-294.

Engle, R.F. (1982), Autoregressive Conditional Heteroscedasticity with Estimates of Variance of U.K. Inflation, *Econometrica*, 50, 987-1008.

Engle, R. F. (1983), Estimates of the Variance of U.S. Inflation Based on the ARCH Model, *Journal of Money Credit and Banking*, 15, 286-301.

Engle, R. F. and Bollerslev, T. (1986), Modelling the Persistence of Conditional Variances, *Econometric Reviews*, 5, 1-50.

Engle, R. F., Lilien, D. M. and Robbins, R. P. (1987), Estimating Time Varying Risk Premia in the Term Structure : the ARCH-M Model, *Econometrica*, 55, 391-408.

Fama, E.F. (1965), The Behaviour of Stock-Market Prices, *Journal of Business*, 38, 34-105.

French, K.R., Schwert, G.W. and Stambough, R.F. (1987), Expected Stock Returns and Volatility, *Journal of Financial Economics*, 19, 3-29.

Friedman, B.M. and Kuttner, K.N. (1988), Time Varying Risk Perceptions and the Pricing of Risky Assets, *Unpublished Manuscript*, Department of Economics, Harvard University, Cambridge, MA.

Gallant, A.R., Rossi, P.E. and Tauchen, G. (1990), Stock Prices and Volume, *Unpublished Manuscript*, Department of Economics, Duke University, Durham, NC.

Geweke, J. (1986), Modeling the Persistence of Conditional Variances : A Comment, *Econometric Reviews*, 5, 57- 61.

Gourieroux, C. (1997), *ARCH Models and Financial Applications*, Springer Series in Statistics.

Higgins, M. L. and Bera, A.K. (1992), A Class of Non-linear ARCH Models, *International Economic Review*, 33, 137-158.

Hsieh, D.A. (1989), Modeling Heteroscedasticity in Daily Foreign Exchange Rates, *Journal of Business and Economic Statistics*, 7, 307-317.

Jorion, P. (1988), On Jump Processes in the Foreign Exchange and Stock Markets, *Review of Financial Studies*, 1, 427- 445.

Malkiel, B. (1979), The Capital Formulation Problem in the United States, *Journal of Finance*, 34, 291-306.

Masry, E. and Tjostheim, D. (1995), Nonparametric Estimation and Identification of Nonlinear ARCH Time Series, *Econometric Theory*, 11, 258-289.

Merton, R.C. (1973), An Intertemporal Capital Asset Pricing Model, *Econometrica*, 41, 867-887.

Merton, R. C. (1980), On Estimating the Expected Return on the Market, *Journal of Financial Economics*, 8, 323-361.

Mills, T. C. (1994), *The Econometric Modelling of Financial Time Series*, Cambridge University Press.

Nelson, D.B. (1991), Conditional Heteroscedasticity in Asset Returns, *Econometrica*, 59, 347-370.

Pagan, A. R. and Schwert, G.W. (1990), Alternative Models for Conditional Stock Volatility, *Journal of Econometrics*, 45, 267-290.

Pantula, S. G. (1986), Modeling the Persistence of Conditional Variances : A Comment, *Econometric Reviews*, 5, 71- 74.

Pantula, S. G. (1985), Estimations of Autoregressive Model with ARCH Errors, *Unpublished Manuscript*, Department of Statistics, North Carolina State University, Raleigh, NC.

Pindyck, R. (1984), Risk, Inflation and the Stock Market, *American Economic Review*, 74, 335-351.

Pindyck, R. (1986), Risk Aversion and Determinants of Stock Market Behevior, *NBER Working Paper No.*1921.

Poterba, J. and Summers, L. (1986), The Persistence of Volatility and Stock Market Fluctuations, *American Economic Review*, 76, 1142-1151.

Weiss, A.A. (1984), ARMA Models with ARCH Errors, *Journal of Time Series Analysis*, 5, 129-143.

10

# A Lingo-Statistical Study of Prose-Literature in Orunodoi Era of Modern Assamese Language

*N.R. Majumdar*

## 1. Introduction

The school of thought of statistical linguistics has immensely contributed towards the development of quantitative methods for studying various problems of linguistic interest. This branch of science deals with statistical techniques to ascertain the basic structure of different languages and their origin and inter-relations to unveil the historical trend in the characteristics of a language due to addition of new words, deletion of old words, generalization of words, changes in phonology and syntax etc., to determine the style of different authors and their differences and resemblances, to depict the chronology of the work of an author, to solve the problem of disputed authorship of valuable manuscripts and documents. Besides these, it contributes to the conformation of a body of conclusions put forth by pure linguist. In the present study an attempt has been made to follow such new avenue of research by considering prose-literature of the Orunodoi era of modern Assamese language as special field.

Assamese in its original form Asamiya is the eastern most Neo-Aryan language spoken by the people native to the modern state of Assam. In formation of the language it also absorbed a good deal of vocabularies as well as certain distinctive peculiarities of grammar and syntax, known to several Tibeto-Burman families of language besides Khasis which belong to Monkhmer family. The dialect, Kamarupa of Magadhi Apabhramsa, developed a language which is known in modern times as Assamese. The first literary period of the modern Assamese is the Orunodoi era which extends from 1846 to 1882. During this period different devoted literati of Assam established the correctness of language, chastity of style, purity of diction and discipline in literature life. The literature of this period is mostly devoted to the subject of religion, natural science and practical knowledge, which render the language a sense of perspective.

## 2. Sampling and Data Collection

A sample of two creative and two constructive prose-texts from the works of this era have been selected through purposive sampling by considering the importance of the work and by observing the due difference to claim and counter-claim numerous contending authors and contesting forms. It has styled those specimens as constructive literature which are non-imaginative, informative and/or critical writings. The samples of texts are :

| *Literature* | *Author* | *Text* |
|---|---|---|
| Creative | 1. Hem Chandra Barua | Bāhire Rangchang Bhitare Koā-Bhāturi |
| | 2. Gunabhiram Barua | Rāmnavami Nātak |
| Constructive | 1. Gunabhiram Barua | Assam Buranji |
| | 2. Gunabhiram Barua | Ānandarām Dhekiāl Phukan |

As a two-stage sampling design under the frame of twenty three pages as first-stage units and five sentences from each page as second-stage units for a text leads to the optimal sampling plan under the joint consideration of cost in terms of time required for the sampling and accuracy in terms of standard error of the estimate (Majumdar, 1986). To estimate the standard errors the technique of interpenetrating sub-samples by dividing the whole sample into five sub-samples have been used. However, during the calculation of average sentence-length, thirty pages in place of twenty three pages as first stage units have been selected in order to achieve a moderate size of sub-sample.

In course of data collection, paragraphing, chapter headings, character headings, sentence written in foreign languages and verse-form etc. have left outside the calculation. A hyphenated word has been considered as a single word. Numerals have been recorded into their word-form equivalents. A sentence which starts on one page and ends in the next page had been decided as a sentence belonging to the next. For estimation of the proportion of parts of speech *Biśesya* and *Sarbanām* (noun and pronoun), *Biśesan* (adjective and adverb), *Kriā* (verb) and abyay (preposition, conjunction and interjection) have, respectively, been grouped as group 1, group 2, group 3 and group 4. Words having 3rd or 5th ease-ending after the stem word on the plane of parts of speech have telescoped as one integral word, and the stem has taken into consideration in deciding its parts of speech. In case of studying the etymological characters of a word, words are categorized into (i) *Tatsama* – word which purely belongs to Sanskrit lexicon, (ii) Native – all words springing from the soil as well as those which are either acclimatized in their original spelling in Sanskrit or developed with the modification of Sanṣkrit spelling by means of linguistic modes of anaptexys or metathesis etc. (iii) Other than Native–all words which are either completely foreign or those borrowed from other Indian/non-Indian languages in modified form, and (iv) Hybrid- all those words which smack of hybridization.

## 3. Estimator and Test-statistic

The estimator is the ratio estimate and is of the form u/v, where u and v are two random variables. But since the standard error of such estimate constructed on the basis of two-stage sampling plan cannot be computed or estimated because of ignorance of population parameters, the procedure of interpenetrating sub-samples of Mahalanobis (1946) and Bhattacharya (1974) had been employed . If $Y_{ij}$

and $X_{ij}$ be the values of characteristics $Y$ and $X$ based on the $j^{th}$ sentence of the $i^{th}$ sub-sample, then the estimators and their standard errors are :

$$\hat{R}_i = \sum_j Y_{ij} / \sum_j X_{ij} \text{ for the } i^{th} \text{ sub-sample}$$

$$\hat{R} = \sum_i \sum_j Y_{ij} / \sum_i \sum_j X_{ij} \text{ for the whole sample}$$

$$\text{and S.E. } (\hat{R}_i) = \left[ \sum_i \left( \hat{R}_i - \hat{R} \right)^2 / k(k-1) \right]^{\frac{1}{2}}$$

where $k$ is the number of sub-samples.For testing the null hypothesis that for a particular language character there is no difference between two texts/two types of literature, the appropriate test-statistics is

$$d = \left[ \hat{R}_i - \hat{R}_2 \right] / \left[ SE(\hat{R}_1)^2 + SE(\hat{R}_2)^2 \right]^{\frac{1}{2}},$$

where $\hat{R}_1$ and $\hat{R}_2$ are the estimates of that character for two comparable texts. Information conveyed by a syllable for communication is measured by the term 'entropy' defined as $H = \sum_i P_i \log_e P_i$ where $P_i$ denotes the proportion of words of length $i$.

## 4. Results and Discussions

The calculated $\hat{R}_i$ 's on the basis of a two-stage random sampling taken from each of the experimental texts and worked out S.E.( $\hat{R}_i$ ) by the principle of interpenetrating sub-samples technique are given in the following table :

| *Type of literature* | | *Creative* | | *Constructive* | |
|---|---|---|---|---|---|
| *Texts* | | *Koā-Bhaturi* $(T_1)$ | *Rāmnavami Nātak* $(T_2)$ | *Assam Buranji* $(T_3)$ | *Ānandram Dhekil Pukhan* $(T_4)$ |
| Kind of literature | | Metaphoric | Drama | History | Biography |
| No. of sample words | | 1744 | 716 | 1512 | 1171 |
| Word length in syllables | | 2.3956 ± 0.0253 | 2.1494 ± 0.0401 | 2.3745 ± 0.0366 | 2.4347 ± 0.0350 |
| Proportion (in %) of Parts of Speech | | | | | |
| $G_1$: | Noun and Pronoun words | 64.27 ± 2.2127 | 55.73 ± 2.2761 | 66.75 ± 2.2279 | 66.69 ± 1.3151 |
| $G_2$: | Adjective and Adverb words | 7.28 ±1.0510 | 9.78 ±1.2535 | 7.03 ± 0.8250 | 5.64 ± 0.3684 |
| $G_3$: | Verb words | 21.38 ±0.6311 | 26.40 ±1.4318 | 17.54 ±1.4020 | 19.13 ±1.1973 |
| $G_4$: | Preposition, Conjunction and Interjection | 7.05 ±1.4378 | 8.10 ±0.9181 | 8.67 ±0.6397 | 8.54 ±0.3471 |
| Proportion (in %) of Etymological Characters | | | | | |
| $C_1$: | Tatsama words | 5.22 ±0.5736 | 11.17 ±2.8405 | 24.70 ±2.5976 | 11.10 ±1.2003 |
| $C_2$: | Native words | 87.31 ±1.7647 | 83.66 ±2.8618 | 64.19 ±4.6759 | 77.88 ±0.8629 |
| $C_3$: | Other than Native words | 6.88 ±1.3543 | 5.03 ±1.6275 | 10.58 ±2.3582 | 11.02 ±0.5332 |
| $C_4$: | Hybrid words | 0.57 ±0.13 | 0.14 ±0.17 | 0.54 ±0.23 | 0.00 ±0.00 |
| | Sentence length | 15.83 ±0.5423 | 6.25 ±0.2223 | 17.79 ±0.8073 | 10.69 ±0.4569 |
| | Entropy per syllable | 2.08 | 2.81 | 1.95 | 2.02 |

The values of 'd' statistic for comparison of two texts at a time with respect to the use of (i) word-length in syllable, (ii) proportion of noun and pronoun, (iii) proportion of adjective and adverb, (iv) proportion of verb, (v) proportion of preposition, conjunction and interjection, (vi) proportion of *Tatsama*, (vii) proportion of Native, (viii) proportion of Other than Native, (ix) Proportion of Hybrid and (x) sentence length in words are estimated and shown in the following tables. One star (*) and two stars (**) indicate the values significant at 5% and 1% levels of significance, respectively.

(i) For word-length

| Text | $T_2$ | $T_3$ | $T_4$ |
|---|---|---|---|
| $T_1$ | 5.19** | 0.47 | 0.91 |
| $T_2$ | - | 4.15** | 5.36* |
| $T_3$ | - | - | 1.19 |

(ii) For noun and pronoun

| Text | $T_2$ | $T_3$ | $T_4$ |
|---|---|---|---|
| $T_1$ | 2.69** | 0.80 | 0.94 |
| $T_2$ | - | 3.46** | 4.18** |
| $T_3$ | - | - | 0.02 |

(iii) For adjective and adverb

| Text | $T_2$ | $T_3$ | $T_4$ |
|---|---|---|---|
| $T_1$ | 1.53 | 0.19 | 1.47 |
| $T_2$ | - | 1.83 | 3.17** |
| $T_3$ | - | - | 1.54 |

(iv) For verb

| Text | $T_2$ | $T_3$ | $T_4$ |
|---|---|---|---|
| $T_1$ | 3.21** | 2.50** | 1.66 |
| $T_2$ | - | 4.92** | 3.90** |
| $T_3$ | - | - | 0.86 |

(v) For preposition, conjunction and interjection

| Text | $T_2$ | $T_3$ | $T_4$ |
|---|---|---|---|
| $T_1$ | 0.60 | 1.03 | 1.01 |
| $T_2$ | - | 0.48 | 0.42 |
| $T_3$ | - | - | 0.18 |

(vi) For Tatsama

| Text | $T_2$ | $T_3$ | $T_4$ |
|---|---|---|---|
| $T_1$ | 2.05* | 7.32** | 4.42** |
| $T_2$ | - | 3.52** | 0.02 |
| $T_3$ | - | - | 4.75** |

(vii) For native

| Text | $T_2$ | $T_3$ | $T_4$ |
|---|---|---|---|
| $T_1$ | 1.09 | 4.63** | 4.80** |
| $T_2$ | - | 3.55** | 1.93 |
| $T_3$ | - | - | 2.88** |

(viii) For other than native

| Text | $T_2$ | $T_3$ | $T_4$ |
|---|---|---|---|
| $T_1$ | 0.87 | 1.36 | 2.84** |
| $T_2$ | - | 1.97 | 3.41** |
| $T_3$ | - | - | 0.18 |

(ix) For hybrid

| Text | $T_2$ | $T_3$ | $T_4$ |
|---|---|---|---|
| $T_1$ | 2.05* | 0.11 | 4.38** |
| $T_2$ | - | 2.35** | 0.82 |
| $T_3$ | - | - | 2.35** |

(x) For sentence-length

| Text | $T_2$ | $T_3$ | $T_4$ |
|---|---|---|---|
| $T_1$ | 27.89** | 2.07* | 10.22** |
| $T_2$ | - | 16.46** | 17.20** |
| $T_3$ | - | - | 7.65** |

Language interpretation of the findings can be drawn as in the following table :

**Structural Differences between Texts**

| *Structure of* | *Differ significantly from that of* | *Due to its formation with Percentage of words* | | | | | | | | | |
|---|---|---|---|---|---|---|---|---|---|---|---|
| | | *Word-length in syllable* | *Nou.: and Pronoun* | *Adjective and Adverb* | *Verb* | *Preposition, Conjunction and Interjection* | *Tatsama* | *Native* | *Other than Native* | *Hybrid* | *Sentence length in words* |
| Bāhire Rangchang Bhitare Koā Bhāturi ($T_1$ : Metaphorical) | Rāmana-vami Nātak ($T_2$ : Drama) | Bigger | More | - | Less | - | Less | - | - | More | Longer |
| | Assam Buranji ($T_3$: History) | - | - | - | Less | - | Less | More | - | - | Shorter |
| | Ānandarām Dhekiāl Phukan ($T_4$: Biography) | - | - | - | - | - | Less | More | Less | More | Longer |
| Rāmnavami Nātak ($T_2$ : Drama) | Assam Buranji ($T_3$: History) | Smaller | Less | - | More | - | Less | More | - | Less | Shorter |
| | Ānandarām Dhekiāl ($T_4$: Biography) | Smaller | Less | More | More | - | - | - | Less | - | Shorter |
| Assam Buranji ($T_3$ : History) | Ānandarām Dhekiāl Phukan ($T_4$: Biography) | - | - | - | - | - | More | Less | - | More | Longer |

## 5. Conclusion

There exists a general conclusion that an author uses different styles for creating different kinds of literature like drama, novel, metaphorical one, biography, historical writing etc. This conclusion put forth by pure literature has been confirmed by this statistical investigation on Gunabhiram Barua, a famous writer of classical temper. It is observed that in depicting history and biography, Barua is seen to adopt a style by taking significantly more *tatsama* words, less native words, more hybrid words and longer sentences in historical piece than those in biographical one.

As in this study, sample of two texts of creative literature and two texts of constructive literature have been selected by purposive way from their respective voluminous inventory and as the estimates of parameters have also been calculated on the basis of sample taken from selected pieces rather than complete enumeration, the estimates of parameters for overall creative, constructive as well as for overall literature of the period, Orunodoi could not be worked out due to the existence of heteroscedasticity problem. However, the present methodological approach may be applicable to the study of language characteristics by taking a sufficiently large sample from the text.

### References

Barua, H.C. (1959), *Bahire Rangchang Bhitare Koā-Bhaturi* (5th Edn.), Hemchandra Puthi Bharal, Sibsagar.

Barua, Gunabhiram (1870), *Ramnavami Natak*, Edited by Jatindra Mohan Bhattacharjee in 1966. Moskhada Pustakalaya, Gauhati.

Barua, Gunabhiram (1875), *Assam Buranji*, Reprinted by Secretary, Publication Board, Assam, Gauhati in 1972.

Barua, Gunabhiram (1880), *Anandaram Dhekial Phukan*, Reprinted by Secretary, Publication Board, Assam, Gauhati in 1971.

Bhattacharaya, N. (1974), A Statistical Study of Word-length in Bengali Prose, *Sankhya, Series B*, 36(4), 323-347.

Mahalanobis, P.C. (1946), Recent Experiments in Statistical Sampling in Indian Statistical Institute, *J. Roy. Stat. Soc.*, 109, 325-378.

Mazumdar, N.R. (1986), *Some Studies on Quantitative Linguistics with a Board Survey on Modern Assamese Literature*, Unpublished Ph. D. Thesis.

11

# Estimation Procedures in Cluster Sampling

***S. J. Amdekar***

## 1. Introduction

In cluster sampling a cluster is the sampling unit and it is generally used when list of all units in the population is not available. Cluster sampling is advantageous from the point of view of operational convenience and reduction in cost. It can also be used in the situations where the list of all units is available. In such cases, clusters are formed either before selecting the sample or after selecting the sample (Goel and Singh, 1973). Moreover, clusters can be overlapping or non-overlapping. Two clusters are said to be overlapping if they have at least one common unit. For example, in surveys with village as a sampling unit a random sample of n villages is selected and to each of the selected village nearby villages lying within some distance are grouped to form clusters. Obviously, clusters so formed will be overlapping. In this paper estimation procedures in overlapping and non-overlapping clusters are described.

## 2. Overlapping Clusters

First consider the case of overlapping clusters. Let there be a population of N units with X and Y as auxiliary and study variables. A reciprocal association scheme for cluster formation is assumed. That is, if unit $U_i$ is selected in the sample and unit $U_j$ becomes member of its cluster, then $U_i$ also becomes member of cluster of $U_j$. Let $M_i$ be the cluster size of $U_i$. That is, if $U_i$ is selected then $(M_i - 1)$ units are associated with it to form a cluster of $M_i$ units. Let an SRSWOR of n units be selected and clusters be formed as explained above. Then an unbiased estimator of population mean is given by $\hat{\bar{y}} = \Sigma T_i / n$, where $T_i = \Sigma y_j / M_j$, the sum is over those $U_j$ which are members of $U_i$. Further, $V(\bar{y}) = (n^{-1} - N^{-1}) \Sigma (T_i - \bar{Y})^2 / (N-1)$. Thus, we have a very simple estimator which requires knowledge of $M_i$ values of only those units which are included in the sample (Amdekar, 1985).

### 3. Use of Auxiliary Variable

From the expression of $T_i$ we note that whenever $U_j$ is included in any cluster weight associated with $Y_j$ is $1/M_j$, i.e., a constant weight for a given $Y_j$. Now we use auxiliary variable to determine the weights so as to minimize $V(\bar{y})$.

Consider clusters of size two only. Thus, the method of cluster formation is as follows. Draw an SRSWOR of n units. If $U_i$ is included, the sample $U_{i+1}$ is associated with it to get a cluster of two units. So here, $T_i = (1 - r_i) Y_i + r_i Y_{i'}$, where $r_i$'s are unknown weights. These are determined such that $T_i = \bar{y}$ for each i. From simple algebra it can be shown that weights r's are given by $r_i = [(i-1)\bar{X} - A_{i-1}] / X_i$, where

$A_i = \sum_{j=1}^{i} X_j$. It can be seen that if these weights are used, $V(\bar{y}) = 0$.

From the formula of $r_i$ we observe that it depends on the arrangement of units in the population. Consequently, performance of the estimator also depends on arrangement of the units. One conservative consideration is to arrange the units so that $r_i$ is in the range $(0,1)$. One possible way to achieve this is to arrange the units as given below :

$$U_{(1)}, U_{(N)}, U_{(2)}, U_{(N-1)}, U_{(3)}, U_{(N-2)}, \dots, U_{(N/2+1)}.$$

## 4. Ratio Estimator in Overlapping Clusters

There is one more way to use auxiliary information in overlapping clusters. In usual cluster sampling it has been shown in the literature that correlation between X and Y increases with the increase in cluster size. Moreover, it has been shown by Mishro and Sukhatme (1972) that even if intraclass correlation is positive, cluster sampling in conjunction with ratio estimator is likely to be more efficient than simple random sampling in conjunction with ratio estimator.

Therefore, in overlapping clusters also one can use ratio estimator. Amdekar (1996) carried out empirical studies and observed that overlapping cluster sampling in conjunction with ratio estimator is also likely to be more efficient than simple random sampling in conjunction with ratio estimator.

## 5. Non-overlapping Clusters

Now consider the case of the non-overlapping clusters. Here also we assume reciprocal association scheme.

Select one unit (called key unit) from the population and form the first cluster of size m by selecting (m-1) out of $M_i$-1 of its associates. Select next unit out of (N-m) units and similarly form the second cluster. Repeat the procedure to get a sample of n clusters.

## 6. Estimation of Population Total

Clusters have been formed successively at each draw. Next

the estimators corresponding to these clusters are developed.. Results are available in Amdekar (1995).

## References

Amdekar, S. J. (1985), An Unbiased Estimator in Overlapping Clusters. *Cal. Stat. Assoc. Bull.*

Amdekar, S. J. (1995), Estimation in Non-overlapping Clusters, *J. Ind. Stat. Assoc.*

Amdekar, S. J. (1996), A Note on Ratio Estimator in Overlapping Clusters, *J. Ind. Soc. of Agri. Stats.*

Goel, B.B.P.S. and Singh, D. (1973), On the Formation of Clusters, *J. Ind. Soc. of Agri. Stats.*

Mishro, G.K. and Sukhatme, B.V. (1972), Efficiency of Cluster Sampling in Conjunction with Ratio and Regression Methods of Estimation, *J. Ind. Soc. of Agri. Stats.*

12

# Application of Statistics in Genetic Analysis of Qualitative and Quantitative Characters

*P. C. Kole*

## 1. Introduction

The success of genetic experiments and breeding depends mainly on the sound genetical investigations carried out in the field and laboratory. Such experiments involve the biological populations to understand the genetics of a trait. The genetic constitution of an individual is known as its genotype and the genotype under the influence of environment results in the expression of a trait, which is known as phenotype. Since the expression of a phenotype is due to interaction between genotype and environment, there is no one-to-one correspondence between genotype and phenotype. Depending upon the degree of influence of environment on the phenotype, characters can be broadly divided into two groups : qualitative and quantitative. Effect of environment and limitations in using large number of individuals in genetic experiments necessitate the application of different statistical and biometrical techniques in genetic experiments.

## 2. Analysis of Qualitative Characters

Qualitative characters are those, which show discontinuous

variation with distinct phenotypic classes. These characters are less influenced by environment and controlled by major gene(s), which are analyzed with the help of counts and ratios. Genetic analysis of qualitative characters is mainly related to the study of inheritance of characters and linkage among genes. In both the cases, chi-square ($\chi^2$) statistics can reliably be used, provided some precautions are taken.

### 2.1 Inheritance of Character

The inheritance of character is related to the detection of number of gene(s) that control the phenotypic expression of the character. Most commonly used test to study the pattern of inheritance of qualitative characters is $\chi^2$ test, whose test-statistic is given by :

$$\chi^2 = \sum_{i=1}^{k} \frac{(O_i - E_i)^2}{E_i} \sim \chi^2_{k-1},$$

where $k$ is the number of classes, and $O_i$ and $E_i$ represent observed and expected frequencies of the $i^{th}$ class, respectively.

Observed data of different classes are tested with chi-square statistic for their fit to an expected ratio. After fitting to a particular ratio, the number of genes (n, say) controlling a character is determined on the basis of the smallest perfect population size ($4^n$) in $F_2$-generation. If the fitted ratio of different phenotypic classes of a character in $F_2$-generation is 3:1, then the smallest perfect population size is $(3+1)^1$, which leads to the conclusion of single gene (monogenic) inheritance of the character. Similarly, if the fitted ratio is 9:3:3:1 or its different modifications (12:3:1, 9:6:1, 9:3:4, 15:1, 13:3, 9:7 etc.), then the smallest perfect population size is $4^2$, which leads to the conclusion of digenic inheritance of the character. Different modifications of a particular ratio having the same size of the smallest perfect population are due to the different modes of interaction between or among genes.

The $\chi^2$ test can be employed when some quantitative character behaves in a manner similar to that of qualitative character. In such cases, optimum environment should be provided for better phenotypic manifestation of the genes. The results may be interpreted with the help of a frequency curve, which will give the idea of major gene(s) plus minor genes.

Parents, hybrids ($F_1$'s) and segregating generation ($F_2$'s) should be grown in the same environment. Validity of the test can be confirmed from the analysis of segregation within the different families of $F_3$ generation. Analysis of segregation in $F_3$-generation helps in identification of the accurate ratio, with which the genes are actually segregating, from among a number of ratios, in $F_2$-generation, all of which give good fit to the observed data. It is perhaps better to compare the uncorrected value of $\chi^2$ with the corrected value following Yates' correction when the degree of freedom (d.f.) is one and expected frequency of any class is between 5 to 10. The $\chi^2$-test is used only on the numerical data, not on the percentages or ratios.

### 2.2 Detection of Linkage

Two genes (say, 'A' and 'B'), when located on a chromosome, may inherit together. Tendency of two or more genes to be inherited together is known as linkage. There are two phases of linkage : coupling and repulsion. In coupling phase, either all dominant genes ('A' and 'B') or all recessive alleles ('a' and 'b') are located on a chromosome. In repulsion phase, one dominant gene is linked with recessive allele of the other gene ('A' with 'b' or 'a' with 'B'). If genotype 'AABB' (having all dominant genes) is crossed with 'aabb' (having all recessive alleles, *i.e.*, double recessive homozygote), which are in coupling phase of linkage, the $F_1$ would be 'AaBb'(dihybrid, i.e., hybrid for two genes). The $F_1$ 'AaBb' produces 'AB' and 'ab' gametes during gametogenesis, which are parental types. In addition, if there is crossing-over between loci 'A' and 'B', there will be the formation of 'Ab' and 'aB'gametes, which will be recombinant or cross-over types. Similarly, if 'AAbb' is crossed with 'aaBB', which are in repulsion phase of linkage, the $F_1$ would be 'AaBb', which will produce 'Ab' and 'aB' gametes (parental types), and 'AB' and 'ab' gametes (recombinant or cross-over types), if there is crossing-over between loci 'A' and 'B', during gametogenesis. When the $F_1$ ('AaBb') is crossed with the parent 'aabb', it is called as back cross or more specifically as test cross.

Two genes when located on two different chromosomes assort independently to give rise to typical dihybrid ratio of 9:3:3:1 in $F_2$-generation, assuming complete dominance at both the loci. Failure to show typical dihybrid ratio of 9:3:3:1 in $F_2$-generation or 1:1:1:1 in test cross is an indication of abnormal segregation at either loci or involvement of linkage. If gene 'A' is segregating in the ratio of $r_a$:1 and 'B' is segregating in the ratio of $r_b$:1, then their joint segregation ratio to four classes would be $r_a r_b : r_a : r_b : 1$ for the phenotypic classes AB : Ab : aB : ab with the corresponding observed frequencies of $O_{ab} : O_a : O_b : O$.

$\chi^2$-value for segregation will be :

(i) A-gene : $$\chi_A^2 = \frac{\left[O_{ab}+O_a-r_a(O_b+0)\right]^2}{n_{ra}}$$

(ii) B-gene : $$\chi_B^2 = \frac{\left[O_{ab}+O_b-r_b(O_a+0)\right]^2}{n_{rb}}$$ with 1 d.f.,

where $n = O_{ab} + O_a + O_b + O$.

The total $\chi_T^2$ with 3 d.f. can be obtained from the four classes using

$$\chi^2 = \sum_{i=1}^{4} \frac{(O_i - E_i)^2}{E_i}.$$

Linkage can be detected through $\chi^2$ linkage ($\chi^2_L$) as

$$\chi^2_L = \chi^2_r - \chi^2_A - \chi^2_B = \frac{[O_{ab}+r_aO_b-r_bO_a+r_ar_bO]^2}{n_{rb}}$$

If $\chi^2_L$ is significant, the results show linkage.

Once linkage between genes is established, next question is what the strength of linkage is. This is measured from recombination value, 'p' which shows inverse relationship with the strength of linkage. The maximum value of 'p' is 0.5. There are four methods to estimate recombination value from $F_2$-data.

(a) *Square root method* (*Kuspira and Bhambani, 1984*)

$$p = 1 - 2\sqrt{DRH} \qquad \text{(coupling phase)}$$

$$p = 2\sqrt{DRH} \qquad \text{(repulsion phase),}$$

where DRH is the frequency of double recessive homozygote, i.e., O/n.

(b) *Emersion's method* (*Emersion, 1916*)

Let $\theta = \dfrac{E-M}{n}$, where $E = O_{ab} + O$, $M = O_a + O_b$.

$$p = 1 - \sqrt{\theta} \qquad \text{(coupling phase)}$$

or, $p = \sqrt{\theta}$ (repulsion phase).

(c) *Product ratio method (Fisher and Balmukund, 1928)*

The ratio of product of crossover to the product of parental types, called product ratio (X), is a function of recombination. Which is given by

$$X = \frac{O_a \times O_b}{O_{ab} \times O} \text{ (coupling phase)}$$

$$= \frac{O_{ab} \times O}{O_a \times O_b} \text{ (repulsion phase)}$$

(d) *Maximum likelihood method (Haldane, 1919)*

Value of θ is calculated using a quadratic equation

$$n\,\theta^2 - \theta\,\{O_{ab} - 2(O_a + O_b) - O\} - 2\,O = 0,$$

$\theta$ is calculated using $x = \dfrac{-b \pm \sqrt{b^2 - 4ac}}{2a}$ in $ax^2 + bx + c = 0$ equation.

Now recombination value is estimated by $p = \sqrt{\theta}$ or $1 - \sqrt{\theta}$.

For linkage analyses large population in segregating generation should be grown so that rare phenotype or genotype can appear in reasonable number. Out of four methods, square root method is very simple, but less reliable. The other three methods give almost similar results. But if crossing-over does not occur in either or both sexes, recombination values cannot be estimated from $F_2$-data by any of the methods.

### 3. Analysis of Quantitative Characters

The majority of agronomic characters show metric variation which are controlled by many genes whose individual effects are small and their segregation along with the effects of environmental variation produces a continuous phenotypic distribution. Genetic analysis of such quantitative characters is accomplished by applying biometric genetical procedures that use analysis of generation means – a first degree statistic, and variances and covariances – second degree statistics. The genetic information obtained from such analyses help breeders in making judicious choice of parents, base population and methods of breeding and selection along with the knowledge of speed and limits to progress.

### 3.1 Analysis of Generation Means

The gene effect can be estimated in the form of additive (d), dominance (h) and epistatic parameters. Based on six populations, $P_1$ (Parent 1), $P_2$ (Parent 2), $F_1$ ($P_1 \times P_2$), $F_2$ (selfing of $F_1$s), $BC_1$ ($F_1 \times P_1$) and $BC_2$ ($F_1 \times P_2$), the various gene effects are as follows :

(i) Mean effect (m) = $\bar{F}_2$

(ii) Additive effect (d) = $\overline{BC}_1 - \overline{BC}_2$

(iii) Dominance effect (h) = $\bar{F}_1 - 4\,\bar{F}_2 - 0.5\,\bar{P}_1 - 0.5\,\bar{P}_2 + 2\,\overline{BC}_1 + 2\overline{BC}_2$.

(iv) Epistatic effect :

(a) Additive × additive effect (i) = $2\,\overline{BC}_1 + 2\overline{BC}_2 - 4\,\bar{F}_2$

(b) Additive × dominance effect (j) = $\overline{BC}_1 - 0.5\,\bar{P}_1 - \overline{BC}_2 + 0.5\,\bar{P}_2$

(c) Dominance × dominance effect (l)

$$= \bar{P}_1 + \bar{P}_2 + 2\bar{F}_1 + 4\bar{F}_2 - 4\,\overline{BC}_1 - 4\,\overline{BC}_2$$

Significance of various gene effects is tested by computing standard errors of the corresponding population means. If one or more estimates are non-significant, re-estimation is done using weighted least square procedure after ignoring the non-significant parameters. This produces an estimate of $\chi^2$ that tests the goodness of fit.

The sign and magnitude of (h) and (l) are not influenced by the degree of dispersion and are used to identify the types of epistasis. A similar sign indicates complementary epistasis while opposite sign indicates duplicate epistasis. Complementary epistasis is expected to magnify variation in $F_2$ but a duplicate type will reduce the genetic variation.

The detection, estimation and interpretation of epistatic parameters are more robust with the first degree statistic and also the type of experiment required is smaller and simpler than the second degree statistic.

### 3.2 Analysis of Variance

Phenotype (P) is the joint action of genotype (G) and environment (E). Phenotypic variance ($\sigma^2_P$) is the total of genotypic variance ($\sigma^2_G$) and environmental variance ($\sigma^2_E$), provided action of 'G' is independent of 'E'. For partitioning of total variance into heritable ($\sigma^2_G$) and nonheritable components ($\sigma^2_E$), trials involving a number of genetically diverse varieties or advanced breeding lines (genotypes) are conducted in randomized block design. The mean squares (MS) from analysis of variance are equated to their expected values. Therefore, error MS = $\sigma^2_E$. Here

Treatment (Genotype) MS = $\sigma^2_E + r\,\sigma^2_G$ ( r = number of replications) and $\sigma^2_P = \sigma^2_G + \sigma^2_E$,

Heritability in broad sense (h²) = $\dfrac{\sigma^2_G}{\sigma^2_P} \times 100\ \%$,

Phenotypic coefficient of variation (PCV) = $\frac{\sigma_P}{\bar{x}} \times 100\ \%$,

Genotypic coefficient of variation (GCV) = $\frac{\sigma_G}{\bar{x}} \times 100\ \%$,

where $\bar{x}'s$ are the respecive means.

Genetic advance (GA) = i $\sigma_p h^2$ ('i' is the selection intensity, the value of which at 5% is 2.06).

The estimates of phenotypic and genotypic coefficients of variation for different characters help in comparing the extent of variation among different characters. High heritability accompanied with high genetic advance indicates the predominance of additive gene effect for expression of the character and selection for such trait would be effective.

However, for estimates of the above genetic parameters, considerably large number of genotypes from diverse sources should be included in the trial and significant mean square due to genotype is a prerequisite.

## References

Bhattacharya, P.K. and Hari Har Ram (2001), Linkage of Genes for Resistance to Yellow Mosaic Virus, Bihar Hairy Caterpillar (Shilosoma Obliqua Walker) and Bacterial Pustules (X. camprestes pv. Glycines) in the Interspecific Crosses of Soybean, *Indian Agriculturist*, 45, 25-31.

Bhattacharya, P.K., Hari Har Ram and Kole, P.C. (1999), Inheritance of Resistance to Yellow Mosaic Virus in Interspecific Crosses of Soybean, *Euphytica*,108, 157-159.

Dash, S.R. and Kole, P.C. (2001), Studies on Variability, Heritability and Genetic Advance in Fenugreek, *Journal of Interacademicia*, 5, 7-10.

Emersion, R.A. (1916), The Calculation of Linkage Intensities, *American Naturalist*, 50, 411-420.

Fisher, R.A. and Balmukund, B. (1928), The Estimation of Linkage from the Offspring of Selfed Heterozygote, *Journal of Genetics*, 20, 79-92.

Haldane, J.B.S. (1919), The Probable Errors of Calculated Linkage

Values and the Most Accurate Method of Determining Gametic from Certain Zygotic Series, *Journal of Genetics*, 8, 291-297.

Hasib, K.M., Ganguli, P.K. and Kole, P.C. (2002), Genetic Analysis of Some Quantitative Characters in Aromatic Rice involving Induced Mutants, *Tropical Agricultural Research and Extension*, *5*, 1-6.

Kuspira, J. and Bhambani, R.N. (1984), A Simple Method of Estimating Recombination Percentage and Linkage Intensities from $F_2$ Data : Example from Triticum Monococcum and Self fertilized Diploid Plant Species, *Theoretical and Applied Genetics*, 68, 61-67.

Panse, V.G. and Sukhatme, P.V. (1985), *Statistical Methods for Agricultural Workers*, Indian Council of Agricultural Research Publications, New Delhi, India.

Singh, R.K. and Chaudhary, B.D. (1985), *Biometrical Methods in Quantitative Genetic Analysis*, Kalyani Publishers, New Delhi, India.

13

# Use of Statistics in Assigning Fuzzy Membership Values to Clay Minerals of Variable Composition

***Chandrika Varadachari and Kunal Ghosh***

## 1. Introduction

Clay minerals are the major inorganic solids on the Earth's surface and constitute the bulk of the soil matrix. At the first level, clay minerals are classified as 1:1, 2:1 and 2:2 depending on the ratio of silica tetrahedral sheet to alumina/magnesia octahedral sheet. The 2:1 clay minerals are again divided into several groups depending on crystallographic and compositional differences. These are clay micas, smectites and vermiculites (Newman and Brown, 1987). There is a further subdivision at the subgroup level, viz., illite, glauconite and celadonite in the clay mica group and montmorillonite, beidellite, nontronite and saponite in the smectite group. Chemical compositions of these clay minerals are not unique but vary over a wide range of values. Such compositional diffuseness makes it difficult to classify clays from chemical composition data and presents inherent difficulties in thermodynamic treatment and modelling of these systems.

Linear discriminant analysis, although popularly used in classification of biological systems (Seber, 1984), has not been much utilized in mineralogical systems. Recently (Varadachari and Mukherjee, 2004) this method was used to analyze compositional differences between clay mineral groups and subgroups and it was shown that clay minerals are chemically distinct and can be classified by compositional data alone. This led to the development of another unique approach for quantifying the compositional vagueness of clay minerals with the application of the concepts of fuzzy logic (Varadachari et al., 2003). Fuzzy logic provides an alternative mathematical approach that is suitable for modelling and digitizing complex situations and is increasingly used in pattern recognition, robotics, image processing, etc., although geochemical applications are limited. By using fuzzy logic in analyzing clay compositions, it was shown (Varadachari et al., 2003) that a clay mineral could be described in terms of how compositionally representative it is, of its own type (belongingness) and to what extent it resembles other types (compositional overlap). This helps to grade such minerals, estimate the quality of a sample and determine the extent of transition to other phases from chemical data alone.

Here, we show how statistical methods have helped us to derive the basic requirements for fuzzy mathematical treatment, viz., delineate the minimum chemical parameters for differentiating between clay groups and subgroups, evaluate the differences between the subgroups, if any, and derive ideal or centric compositions to fix 0 and 1 levels.

## 2. Data Set

Chemical composition data of 464 clay minerals collected from the literature scanning over 40 years are available in Varadachari et al. (2003). These include the 2:1 clay mineral groups, viz., smectite, clay mica and vermiculite with the subgroups, montmorillonite, nontronite, beidellite and saponite (in the smectite group), illite, glauconite and celadonite (in the clay mica group) and vermiculite (in the vermiculite group).

## 3. Selection of Parameters

To select the parameters that could be used to differentiate various groups and subgroups of clay minerals, we first studied the significance of the differences between their major chemical parameters, viz., total layer charge (TLC), tetrahedral charge (TC), octahedral charge (OC), total $K^+$, octahedral $Al^{3+}$ [Al (VI)], total $Fe^{3+}$ and octahedral $Mg^{2+}$ [Mg (VI)]. Analysis of the critical differences (1%) revealed (Table 13.1) that the clay micas (illites, glauconites and celadonites), have significantly higher TLC than the smectites (montmorillonite, nontronite, beidellite and saponite). K levels are also significantly different between minerals of these broad groups. No other parameter shows such unequivocal differences. Thus, celadonite and nontronite have similar levels of Al (IV) as do illite and beidellite. Similarly, Mg (VI) and OC levels in illite cannot be differentiated from beidellite or nontronite. Montmorillonites and celadonites have similar TC as do illite and saponite. Therefore, the two parameters that may be used for group level classification are TLC and K contents (Varadachari and Mukherjee, 2004).

When we look at the subgroup level, we find that the minerals cannot be distinguished on the basis of any single parameter. To look further into this problem of whether there are any significant compositional differences between the subgroups, linear discriminant analysis (Seber, 1984) was done.

## 4. Analysis of Classification by Linear Discriminant Analysis

Statistical analysis showed (Varadachari and Mukherjee, 2004) that classification using all 6 parameters, TLC, K, Al (VI), Mg (VI), OC and TC into 8 derived classes produced significant differences. Squared Mahalanobis distances are all highly significant (0.0001%). Experimental groupings when compared with statistical classification showed a good matching with 88% fit of the data.

A two-level classification was then attempted. At the group level, classification was done using two parameters, TLC and K. The three groups derived showed 87.7 % fit between observed and predicted classifications. Squared Mahalanobis distances are also highly significant (0.0001%).

For subgroup level classification, 4 parameters were used, viz., Al (VI), Mg (VI), OC and TC. All 8 subgroups are observed to be clearly and significantly different from each other. Matching between experimental and predicted classifications is around 85-95%. Thus, illites, celadonites and glauconites are chemically distinguished as also are montmorillonites, nontronites, beidellites and saponites. Discriminant analysis has, therefore, helped to resolve an age-old dispute on whether clay minerals are chemically distinct or whether they form a continuous series of solid solutions with no marked differences. It has also helped to statistically arrive at the parameters for defining the clay mineral compositions, which is needed for further work on fuzzy description of their chemical compositions.

## 5. Derivations of Centric Compositions

Clay minerals have no ideal or centric compositions, which are necessary for assignment of fuzzy membership values of 1. Therefore, centric values have to be derived. The procedure adopted was as follows : first, the four parameters which are needed to define subgroups were taken and histograms of the distributions for each mineral subgroup were constructed. Then, outlier values were rejected to improve the robustness of the data. Means of the remaining data were evaluated. The centric value of that parameter for a particular mineral type (Varadachari et al., 2003) was thus obtained. These centric values represent the values that are assigned grade 1 (membership value 1) for that parameter in that subgroup. Thus, a value of 0.63 for K level is assigned grade 1 for the illite subgroup whereas for the montmorillonite subgroup K level grade 1 corresponds to 0.178. Centric values for other chemical constituents such as $Ca^{2+}$, $Na^{+}$, $Fe^{2+}$, etc., were also obtained in this manner.

The procedure was changed at the broad group level due to differences in the number of samples within each constituent subgroup. Here, outliers could not be rejected because by rejecting outliers, we could be rejecting bulk of the data from smaller subgroups. Therefore, in this case, averages of the centric values of constituent subgroups were computed to provide the group centric value for that parameter.

Finally, the centric data were combined to obtain a centric or ideal composition of the minerals. Such centric compositions for the mineral subgroups are shown in Table 13.1.

**Table 13.1 :** Chemical parameter analysis and derived centric formula for clay minerals

| *Parameters* | *Significance of differences at p=1%** | *Mineral* | *Derived centric formula* |
|---|---|---|---|
| TLC | C G V I N M S B | Montmorillonite (M) | $K_{0.178}Na_{0.112}Ca_{0.0745}(Si_{3.823}Al_{0.177})Al_{1.453}Fe^{III}_{0.245}Mg_{0.322}O_{10}(OH)_2$ |
| K | V S B N M I G C | Nontronite (N) | $K_{0.105}Na_{0.107}Ca_{0.125}(Si_{3.592}Al_{0.408})\ Al_{0.052}Fe^{III}_{1.75}Mg_{0.27}O_{10}(OH)_2$ |
| A1 (VI) | S N C V G M I B | Saponite (S) | $K_{0.017}Na_{0.027}Ca_{0.1895}(Si_{3.518}Al_{0.482})Al_{0.047}Fe^{III}_{0.083}Fe^{II}_{0.1105}Mg_{2.724}O_{10}(OH)_2$ |
| Mg (VI) | B I N M G C V S | Beidellite (B) | $K_{0.1328}Na_{0.0632}Ca_{0.1184}(Si_{3.625}Al_{0.375})Al_{1.86}Fe^{III}_{0.054}\ Fe^{II}_{0.0061}Mg_{0.094}O_{10}(OH)_2$ |
| OC | C G M I N B S V | Illite (I) | $K_{0.63}Na_{0.0624}(Si_{3.447}Al_{0.553})Al_{1.685}Fe^{III}_{0.1142}\ Fe^{II}_{0.0247}Mg_{0.2068}O_{10}(OH)_2$ |
| TC | V I S N G B M C | Glauconite (G) | $K_{0.748}Na_{0.05}Ca_{0.03}(Si_{3.65}Al_{0.35})Al_{0.502}Fe^{III}_{0.95}Fe^{II}_{0.148}Mg_{0.42}O_{10}(OH)_2$ |
| | | Celadonite (C) | $K_{0.701}Na_{0.0755}Ca_{0.052}(Si_{3.867}Al_{0.133})\ Al_{0.179}Fe^{III}_{1.0105}\ Fe^{II}_{0.196}Mg_{0.646}O_{10}(OH)_2$ |

TLC : Total layer charge, TC : tetrahedral charge, OC : octahedral charge

* minerals connected by lines (underlined or overlined) do not differ significantly.

## 6. Outer Limits of Composition

It is necessary to define the outer limits of each compositional parameter in order to assign grade value 0 to that chemical parameter within a group or subgroup. Since the data distributions for the parameters are skewed, the centric value is not necessarily equidistant from the two extreme data points. Here, the data point at the maximum distance from the centric value has been assigned grade 0. The scale for grade 0 and 1 is, thus, different for different parameters and different minerals. The maximum observed distance from the centric value, of the $\alpha_i$ parameter, in the mineral group/ subgroup G is given by $\Gamma$. Thus,

$$\Gamma = \max(|\alpha_{i\ centric}^{G} - \alpha_i(s_1)|, |\alpha_{i\ centric}^{G} - \alpha_i(s_2)|, \ldots, |\alpha_{i\ centric}^{G} - \alpha_i(s_r)|),$$

where $\alpha_i(s)$ is a value of a parameter, $\alpha_i$ (e.g., tetrahedral charge) of a sample s in some mineral group G (e.g., montmorillonite) having a database of r samples, $\alpha_{i\ centric}^{G}$ is the centric or typical value of the parameter $\alpha_i$ for the group G. Other $\Gamma$-values for clay subgroups are available in Varadachari et al. (2003).

Thus, we have all parametric values needed to derive fuzzy compositional grades of the clay minerals using the two equations (Varadachari et al., 2003)

$$\mu_{\alpha i}^{G}(s) = 1 - \{|\alpha_i(s) - \alpha_{i\ centric}^{G}| / \Gamma\}$$

$$\mu_{\alpha 1,\alpha 2}^{G} = t(\mu_{\alpha 1}^{G}, \mu_{\alpha 2}^{G}) = \mu_{\alpha 1}^{G} \times \mu_{\alpha 2}^{G} / [(\mu_{\alpha 1}^{G})^{0.5} + (\mu_{\alpha 2}^{G})^{0.5} - (\mu_{\alpha 1}^{G} \times \mu_{\alpha 2}^{G})^{0.5}]^{1/0.5}.$$

### Acknowledgements

The authors are grateful to the Department of Science and Technology, Government of India, New Delhi for financial support. They are also grateful to Professor J.K. Ghosh,

Professor M.K. Chakraborty, Smt. G. Mukherjee and Shri D.P. Goswami for academic help.

## References

Newman, A.C.D. and Brown, G. (1987), The *Chemical Constitution of Clays*, Newman, A.C.D.

Seber, G.A.F. (1984), *Multivariate Observations*, John Wiley, New York, 279-346.

Varadachari, C., Mukherjee, G., Goswami, D.P. and Chakraborty, M.K. (2003), Understanding Clay Minerals with Fuzzy Mathematics, *Naturwissenschaften*, 90, 44-48.

Varadachari, C. and Mukherjee, G. (2004), Discriminant Analysis of Clay Mineral Compositions, *Clays and Clays Minerals*, 52, 311-320.

## 14

# Weather Based Forecasting of Crop Yields, Pests and Diseases— IASRI Models

***Ranjana Agrawal***

### 1. Introduction

Weather plays an important role in crop growth as well as development of pests and diseases. Therefore, models based on weather parameters could provide reliable forecast of crop yield in advance of harvest and also forewarning of pest and disease attack so that suitable plant protection measures could be taken up timely to protect the crops. Various organizations in India and abroad are engaged in developing methodologies for this purpose. A number of techniques have been developed at Indian Agricultural Statistics Research Institute (IASRI) for various crops. In the present paper, some of the models are discussed.

### 2. Yield Forecast Models

Models utilize data on yield and weather variables for the location under consideration for past several years. Crop yield in different years is affected by technological change and weather variability. It can be assumed that the technological factors will increase yield smoothly through time and, therefore, years or some other function of time can be used to study the overall effect of technology on yield. Weather variability, both within and between seasons, is the second and the only uncontrollable source of variability in yields. Weather variables affect the crop differently during different stages of development. Thus extent of weather influence on crop yield depends not only on the magnitude of weather variables but also on the distribution pattern of weather over the crop season which, as such, calls for the necessity of dividing the whole crop season into fine intervals. This will increase number of variables in the model, and, in turn, a large number of constants will have to be evaluated from the data. This will require a long series of data for precise estimation of the constants which may not be available in practice. Thus, a technique based on relatively smaller number of manageable parameters and at the same time taking care of entire weather distribution may solve the problem.

### 2.1 Model Using Composite Weather Variables

Fisher (1924) and Hendricks and Scholl (1943) suggested models which require small number of parameters to be estimated while taking care of distribution pattern of weather over the crop season.

Fisher utilized weekly weather data. He assumed that the effect of change in weather variable in successive weeks would not be abrupt or erratic but an orderly one that follows some mathematical law. He assumed that these effects as well as magnitude of the variable in successive weeks are composed of the terms of a polynomial function of time. Substituting these in usual regression equation,

$$Y = A_0 + A_1X_1 + A_2X_2 + ... + A_nX_n,$$

where Y denotes yield and $X_w$ denotes rainfall in $w^{th}$ week, w = 1, 2,...,n. Utilizing the properties of orthogonal and normalized functions, he obtained

$$Y = A_0 + a_0\rho_0 + a_1\rho_1 + a_2\rho_2 + \ldots + a_k\rho_k,$$

where $A_0, a_0, a_1, a_2, \ldots, a_k$ are constants to be determined and $\rho_i$ (i=1,...k) are distribution constants of $X_w$. Fisher suggested to use k = 5 for most of the practical situations. In fitting this equation for k = 5, the number of constants to be evaluated will remain 7, no matter how finely growing season is divided. This model was used by Fisher (1924) for studying the influence of rainfall on the yield of wheat.

Hendricks and Scholl (1943)modified Fisher's technique. They assumed that a second degree polynomial in week number would be sufficiently flexible to express the effects in successive weeks. Under this assumption, the model suggested by Hendricks and Scholl is

$$Y = A_0 + a_0 \sum_{w=1}^{n} X_w + a_1 \sum_{w=1}^{n} w X_w + a_2 \sum_{w=1}^{n} w^2 X_w .$$

Here $X_w$ = value of weather variable under study in $w^{th}$ week, n = no. of weeks in the crop season, $A_0, a_0, a_1, a_2$ are constants. In this model number of constants to be determined reduces to 4, irrespective of n.

This model was extended to study joint effects of weather variables and an additional variate T representing the year was included to make allowance for time trend.

At Indian Agricultural Statistics Research Institute, the model suggested by Hendricks and Scholl has been modified (Agrawal et al., 1980, 1983; Jain et al., 1980) by expressing the effects of changes in weather variables on yield in the $w^{th}$ week as second degree polynomial in respective correlation coefficients between yield and weather variables. This will explain the relationship in a better way as it gives appropriate weightage to weather in different periods. Under this assumption, the models were developed for studying the effects of weather variables on yield using complete crop season data whereas forecast model utilized partial crop season data. These models were found to be better than the one suggested by Hendricks and Scholl (1943).

The models were further modified (Agrawal et al., 1986) by expressing the effects of changes in weather variables on yield in $w^{th}$ week as a linear function of respective correlation coefficients between yield and weather variables. As trend effect on yield was found to be significant, its effect was removed from yield while calculating correlation coefficients of yield with weather variables to be used as weights. Effects of second degree terms of weather variables were also studied. The results indicated that (i) the models using correlation coefficients based on yield adjusted for trend effect were better than the ones using simple correlations, (ii) inclusion of quadratic terms of weather variables and also the second power of correlation coefficients did not improve the model. This suggests that the following type of model can be used to study effects of weather on yield :

$$Y = A_0 + a_0 Z_0 + a_1 Z_1 + cT + e,$$

where $Z_j = \sum_{w=1}^{n} r_w^j X_w \; ; \quad j = 0, 1.$

Here Y is yield, $r_w$ is correlation coefficient between weather variable in $w^{th}$ week ($X_w$) and yield (adjusted for trend effect) and e is error term. The model was further extended for studying joint effects.

The forecast model has been developed using partial crop season data considering all weather variables simultaneously. The model finally recommended was of the form

$$Y = A_0 + \sum_{i=1}^{p} \sum_{j=0}^{1} a_{ij} Z_{ij} + \sum_{i \neq i'=1}^{p} \sum_{j=0}^{1} a_{ii'j} Z_{ii'j} + cT + e,$$

where

$$Z_{ij} = \sum_{w=1}^{m} r_{iw}^{j} X_{iw} \text{ and } Z_{ii'j} = \sum_{w=1}^{m} r^{j}_{ii'w} X_{iw} X_{i'w},$$

$r_{iw}(r_{ii'w})$ is correlation coefficient between yield (adjusted for trend effect) with $i^{th}$ weather variable(product of $i^{th}$ and $i'^{th}$ weather variable in $w^{th}$ week), m is week of forecast and p is the number of weather variables used. Data starting a fortnight before sowing were considered.Stepwise regression technique was used to select the important composite variables.

These models were used to forecast yield of rice and wheat at district level in different situations, viz. (i) rainfed district having deficient rainfall (rice), (ii) rainfed district having adequate rainfall (rice) and (iii) irrigated district (wheat). The results revealed that reliable forecasts can be obtained using this approach when the crops are 10-12 weeks old, i.e., about 2 months before harvest. This approach was also used to develop forecast model for sugarcane (long duration crop) for Kohlapur district (Mehta et al., 2000). All the coefficients of determination were highly significant. Deviations of forecasts of subsequent years (not included in model development) from observed ones ranged between 5 % and 10 %.

These studies, carried out at district level, revealed data requirement of 20-25 years for reliable forecast. Such a long series may not be available for most of the locations. Therefore, a study has been undertaken to develop the model on agro-climatic zone basis by pooling the data of various districts within the zone so that a long series could be obtained in a relatively shorter period. Models were developed for wheat in Vindhya Plateau zone and for rice in Chhattisgarh Plain and Bastar Plateau zone taken together (as a portion of Bastar district falls under Chhattisgarh Plain zone, whereas remaining under Bastar Plateau zone, and yield figures are available at district level only). Agricultural inputs, previous year's yield and moving averages of yield were taken as the variables taking care of variation between districts within the zone. Year variable was included to take care of technological changes. Different strategies for pooling district level data for the zone were adopted. Results revealed that the reliable forecasts can be obtained by using this methodology at 12 weeks after sowing, i.e., about 2 months before harvest at zone level also. The data requirement reduced to 10-15 years as against 20-25 years for district level models. The study also revealed that forecast model will be appropriate to forecast the yield of zone, even if the data for some districts within the zone are not available at model development stage or at forecasting stage (Agrawal et al., 2001).

Forecasts of subsequent years (not included in model development) are given in Figs.14.1 and 14.2.

### 2.2 Models using Discriminant Function Analysis

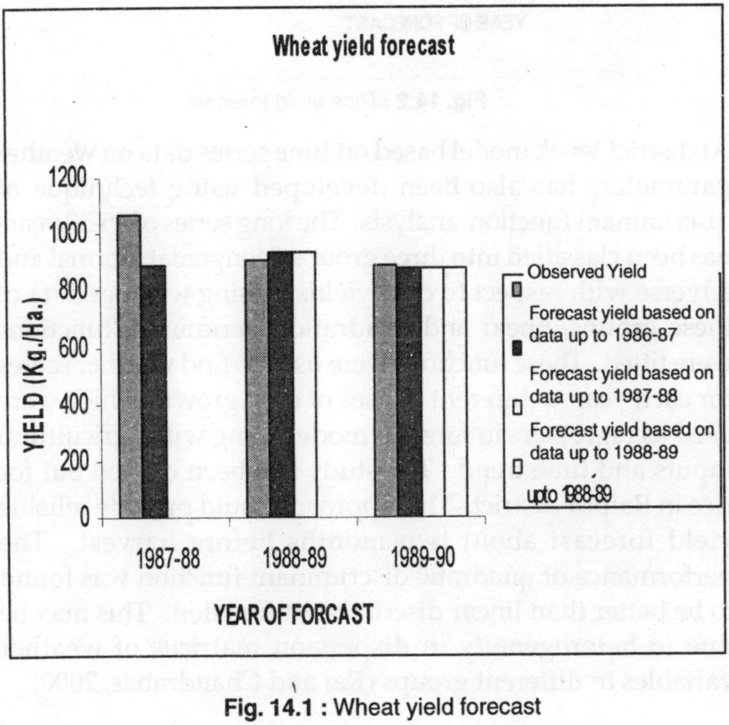


**Fig. 14.1** : Wheat yield forecast

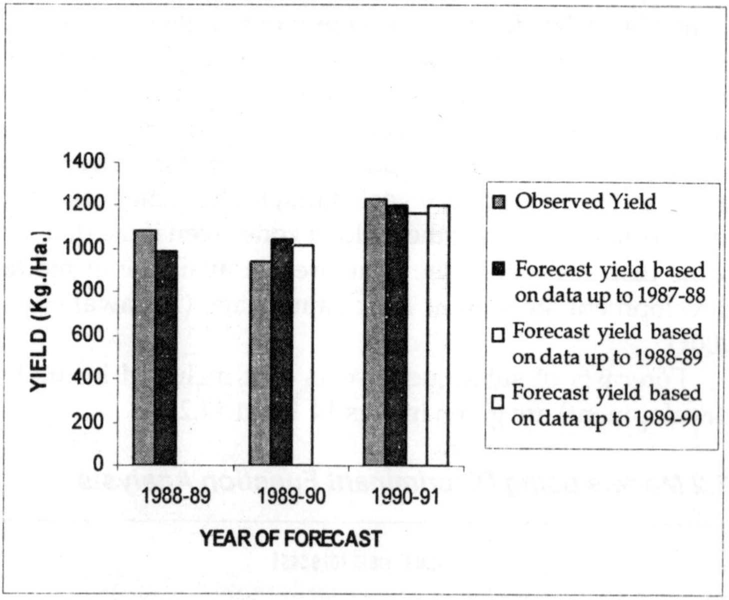


**Fig. 14.2 :** Rice yield forecast

At district level, model based on time series data on weather parameters has also been developed using technique of discriminant function analysis. The long series of 25-30 years has been classified into three groups - congenial, normal and adverse with respect to crop yields. Using weather data of these groups, linear and quadratic discriminant functions were fitted. These functions were used to find weather scores for each year at different phases of crop growth which were used as regressors in forecast model along with agricultural inputs and time trend. The study has been carried out for rice in Raipur district. This approach could provide reliable yield forecast about two months before harvest. The performance of quadratic discriminant function was found to be better than linear discriminant function. This may be due to heterogeneity in dispersion matrices of weather variables in different groups (Rai and Chandrahas, 2000).

### 2.3 Models using Water Balance Technique

Using water balance technique, models have been developed for rainfed crops, rice – Raipur, sorghum – Delhi and Parbhani and maize – Delhi. In this approach, water or deficit or surplus has been worked out at weekly intervals. Weighted stress indices were prepared phase-wise by applying appropriate weights to surplus as well as deficit depending upon stage at which it occurred. These stress indices for each year have been used as regressors along with time trend to develop the forecast model.

The estimated soil moisture in the root zone at the end of i$^{th}$ week was obtained from

$$S_i = S_{i-1} + R_i - WR_i,$$

where $R_i$ is rainfall in the i$^{th}$ week.

$WR_i$ is water requirement of the crop in i$^{th}$ week, which equals $k_i . evap_i$ ($k_i$ is crop coefficient and $evap_i$ is pan evaporation in the i$^{th}$ week).
Stress due to deficit,

$$St_i = 1 - AE_i / WR_i.$$

where $AE_i$ is actual evapotranspiration.
Stress due to surplus was taken as '1' if $S_i$ > water holding capacity.

This technique provided forecast 6 weeks before harvest for sorghum, 4 weeks before harvest for maize and 5 weeks before harvest for rice. Deviations of forecasts from observed yields of subsequent years ranged between 1% and 11% (Saxena et al., 2001).

## 3. Forewarning of Pests and Diseases

Various types of models were developed for different pests and diseases in different situations.

### 3.1 Forewarning System for Aphid

Models were developed for forewarning aphid population at different weeks during the crop season for mustard and potato. Using weekly weather parameters lagged by 1 – 2 weeks, two types of models were developed:

(i) $Y = f(X) + e$

Three types of functions were considered:

cos
log
exponential

(ii) Complex polynomial through GMDH technique :

$$Y = a + \sum_{i=1}^{m} b_i X_i + \sum_{i=1}^{m} \sum_{j=1}^{m} c_{ij} X_i X_j + \sum_{i=1}^{m} \sum_{j=1}^{m} \sum_{k=1}^{m} d_{ijk} X_i X_j X_k + \ldots,$$

$Y$= aphid population of the current week, and
$X_1$'s are weather variables lagged by 1-2 weeks (Trivedi et al.).
As an example, first type of model for 2nd week of January for potato aphid (Pantnagar) was

$$Y = 231.85 - 206.21 \cos(6.37 X_{12} + 13.80) - 199.02 \cos(-22.54 X_{22} + 156.46) - 186.62 \cos(8.83 X_{32} - 228.80),$$

where

$X_{12}$ = maximum temperature in 4th week of December
$X_{22}$ = minimum temperature in 4th week of December
$X_{32}$ = relative humidity in 4th week of December.

The observed and predicted values are given in Fig. 14.3.

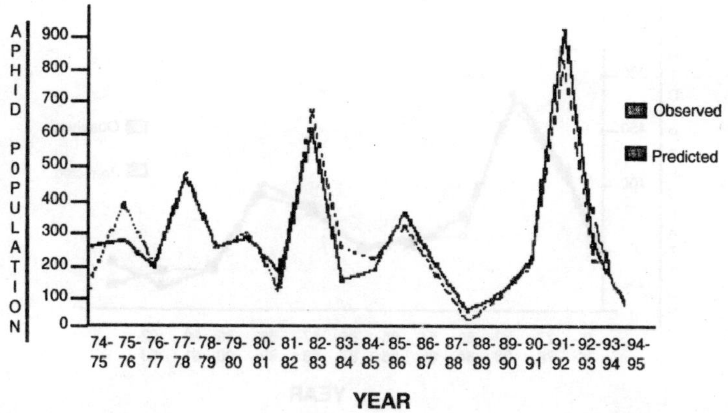


**Fig. 14.3 :** Comparison of observed and predicted aphid population during second week of January at Pantnagar.

One example of complex polynomial for forewarning aphid population (potato – Pantnagar) in 3rd week of December was

$$Y = 18.40 + 0.009\ G_{22}.G_{23}\ \ldots\ (R^2 = 0.96),$$

where

$G_{22} = -370.61 + 329.80\ G_{11} - 207.27\ G_{14} - 28.19\ G^2_{11} - 0.005\ G^2_{14} + 20.47\ G_{11}.G_{14}$

$G_{23} = -76209.00 + 8973.30\ G_{13} - 1569.68\ G_{12} - 249.88\ G^2_{13} - 3.42\ G^2_{12} + 6.23\ G_{13}.G_{12}$,

$G_{11} = X_{21}$ : minimum temperature in 2nd week of December.

$G_{14} = X_{22}$ : minimum temperature in 1st week of December.

$G_{13} = X_{12}$ : maximum temperature in 1st week of December.

$G_{12} = X_{31}$ : relative humidity in 2nd week of December.

Observed and predicted values from this model are given in Fig. 14.4.

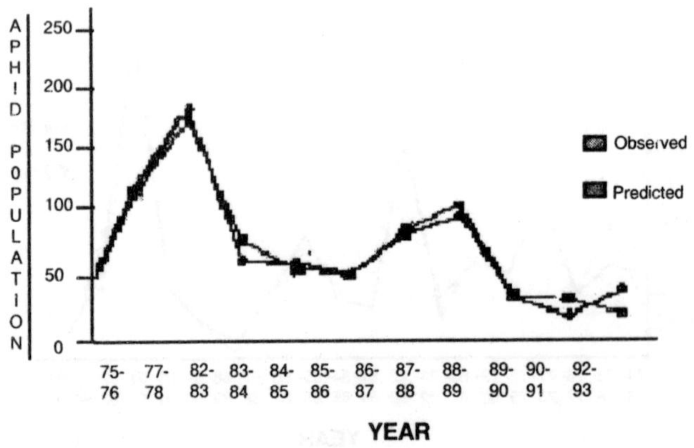


**Fig. 14.4 :** Comparison of observed and predicted aphid population during third week of December at Pantnagar.

### *3.2 Forewarning of Outbreak of Mango Fruitfly*

This study has been taken up at Rehman Khera Farm, Central Institute for Subtropical Horticulture, Lucknow. The variable to forecast was weekly fruitfly population (Number/Trap/Week). The forewarning model was developed on the basis of 6 years' data. As the data were not adequate for development of week-wise models separately, combined model was developed. It was assumed that the fruitfly population in any week will be composed of two components, one due to natural cycle of fruitfly population development and the second, effects of weather fluctuations. Accordingly, the model was developed in two stages. Natural pattern of fruitfly was obtained by fitting proper non-linear function between weekly fruitfly population (averaged over years) and week number. The deviations of individual year's populations in different weeks from natural pattern were related to weather variables (deviations from average) in successive weeks using appropriate functions (Mehta et al., 2001).

The model for natural pattern was

$$Y_t = \frac{33.64 - 1.79t}{1 - 0.16t + 0.0067t^2},$$

$t$ = week number

$Y_t$ = fruitfly population at week $t$.

The observed values and predicted curve are given in Fig.14.5.

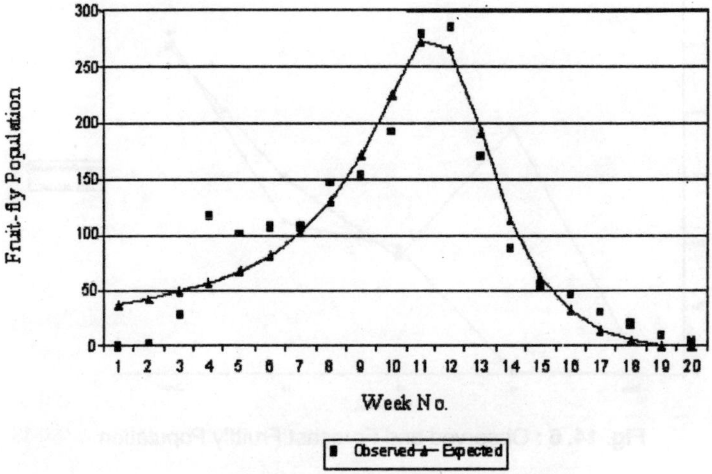


**Fig. 14. 5 :** Natural Pattern of Fruitfly

The final forecast model was

$$Y = -125.766 + 0.665\,(Y_2) + 0.115\,(1/X^2_{22}) + 10.658\,(X^2_{12}) + 0.0013\,(Y^2_3) + 31.788\,(1/Y_3) - 21.317\,(X_{12}) - 2.149\,(1/X^2_{33}) - 1.746\,(1/X^2_{34})$$

$Y$ = deviation of fruitfly population from natural cycle

$Y_2$ = fruitfly population in the 2[nd] lag week

$Y_3$ = fruitfly population in the 3[rd] lag week

$X_{12}$ = maximum temperature in 2[nd] lag week (deviation from average)

$X_{22}$= minimum temperature in 2[nd] lag week (deviation from average)

$X_{33}$= relative humidity in 3[rd] lag week (deviation from average)

$X_{34}$= relative humidity in 4[th] lag week (deviation from average).

Using this model, forewarning can be issued one week in advance. Forewarnings for subsequent year are given in Fig. 14.6.

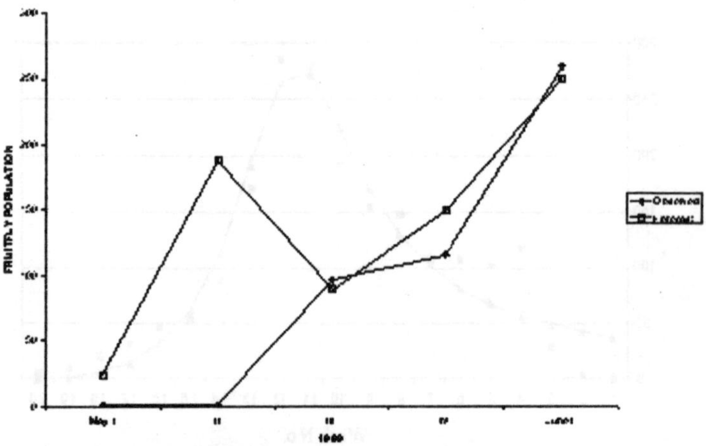


**Fig. 14. 6 : Observed and Forecast Fruitfly Population**

## 3.3 *Forewarning of Powdery Mildew – Mango*

The data were in dichotomous form (1 for presence of disease and 0, otherwise). Logistic model has been developed for Kakori and Malihabad mango belt of U.P.

*The Model*

$$P(Y = 1) = 1 / \{1 + \exp(-z)\} + e,$$

where

$z = b_0 + b_1 X_1 + b_2 X_2$
Y = status of occurrence of powdery mildew
$X_1$ = max. T (1 week lag)
$X_2$ = mean RH (1 week lag).

The model was developed for epidemic status in 3rd week of March. The forewarning can be issued from this model one week in advance by using following decision rule (Ramasubramanian et al., 2002) :

If $p \leq 0.4$ : probability of occurrence of epidemic is minimal
If $0.4 < p < 0.6$ : indecisive zone
If $p \geq 0.6$ : more chance of occurrence of epidemic.

Table 14.1 gives the forewarnings issued by this model and actual observed status of the disease.

**Table 14.1 : Forewarning of Powdery Mildew - Mango**

| *Model based on* | *Forewarning for year* | *Observed epidemic status* | *Forewarning probability* |
|---|---|---|---|
| 1987-95 | 1996 | 1 | 0.66 |
| 1987-95 | 1997 | 0 | 0.52 |
| 1987-95 | 2000 | 0 | 0.34 |
| 1987-96 | 1997 | 0 | 0.35 |
| 1987-96 | 2000 | 0 | 0.34 |
| 1987-97 | 2000 | 0 | 0.29 |

It can be seen from Table 14.1 that there is good matching between forewarning and observed epidemic status.

### *3.4 Weather-based Forewarning of Pests and Diseases – NATP Project*

Under National Agricultural Technology Project funded by World Bank, a project on 'Development of Weather Based Forewarning Systems for Crop Pests and Diseases' has been taken up aiming at developing the database on major pests and diseases and corresponding weather parameters for important crops in various agroclimatic zones and to develop forewarning systems for various pests and diseases. The lead centre of the project is Central Research Institute for Dryland Agriculture, Hyderabad. Indian Agricultural Statistics Research Institute is collaborating in the project with the responsibility of developing statistical models for the purpose. Models under this project have been developed for forewarning Alternaria Blight, White Rust, Powdery Mildew, Downy Mildew and aphid (mustard), Pink boll worm and Whitefly (Cotton), Spodoptera Litura (Groundnut), Pyrilla, Early shoot borer and Top borer (Sugarcane) and Pod fly, Pod borer, MOSIC and Phytophthora Blight (Pigeon pea). (Ramkrishna et al., 2003).

The data consisted of weekly data on pest and disease or data at one point of time during the crop season or in qualitative form for the past years. Experiments were also conducted taking different dates of sowing.

For quantitative data, models were developed for forewarning maximum disease severity and pest population, time of first appearance of diseases/pest, time of maximum disease severity/pest population and age-wise/standard weekwise or yearwise disease severity and pest population. The general form of the model was

$$Y = a_0 + \sum_{i=1}^{p} a_i z_i \sum_{i(\neq j)=1}^{p} b_{ij} z_{ij} + e,$$

where

$$Z_i = \sum_{w=n_1}^{n_2} r_{iw} X_{iw}$$

$$Z_{ij} = \sum_{w=n_1}^{n_2} r_{ijw} X_{iw} X_{jw},$$

$Y$ : variable to forecast

$X_{iw}$: value of i[th] weather parameter in w[th] week

$r_{iw}$ : correlation coefficient between Y and i[th] weather parameter in w[th] week

$r_{ijw}$ : correlation coefficient between Y and product of $x_i$ and $x_j$ in w[th] week

$p$ : number of weather variables,

$e$ : error term

$n_1$ and $n_2$ : initial and final weeks for which weather data were included in the model

In some cases previous disease incidence/pest population (or their indices) and/or previous year's last population has also been included in the model.

In some cases, modified form of the model using unweighted accumulations along with weighted accumulations (of weather parameters) were attempted. The modified form of the model was

$$Y = a_o + \sum_{i=1}^{p}\sum_{j=0}^{1} a_{ij} z_{ji} + \sum_{i(\neq j)=1}^{p}\sum_{j=0}^{1} b_{iij} z_{ii'j} + e,$$

where

$$z_{ij} = \sum_{w=n_1}^{n_2} r_{iw}^{j} x_{iw'}$$

$$z_{ij} = \sum_{w=n_1}^{n_2} r_{ii'w}^{j} x_{iw} x_{i'w}$$

$r_{ii'w}$ is correlation coefficient between Y and product of $X_i$ and $X_{i'}$ in w[th] week, other symbols have same meaning as in the earlier model.

Stepwise regression technique was used for selecting important variables to be included in the model.

In general, the models fitted well for all the available data with all the coefficients of determination highly significant.

In some cases, the modified model performed better whereas in most of the cases, the earlier models worked well or both the models were same. However, for completeness, the modified form of the model is recom-mended. Using these models, reliable forewarnings are possible at least one week in advance. As an example, the forecasts of mustard aphid using modified model are given in Figs. 14.7 -14.9 (Agrawal et al., 2004 ).

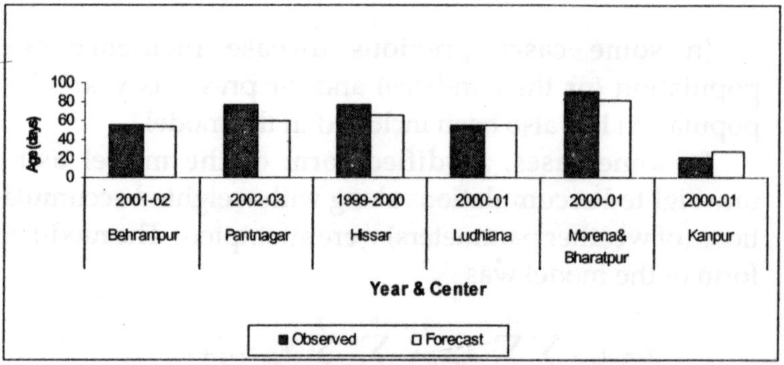


**Fig. 14. 7 :** Crop age at first appearance of aphid

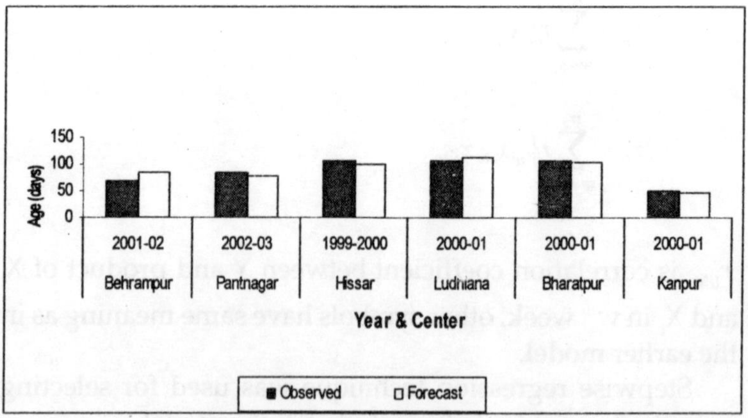


**Fig. 14. 8 :** Crop age at max population

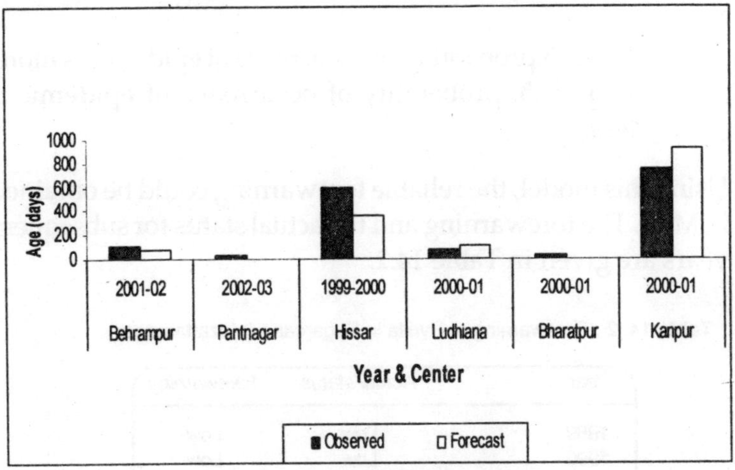


**Fig. 14. 9 : Max Aphid Population**

These forecasts utilized weather data up to 2nd week of December.

For qualitative data, ordinal logistic model was used. This type of model was used for forewarning Pyrilla in sugarcane in Muzaffarnagar district. Here the data was in heterogenous form. In some years, monthly pest population was given, in some years data at one point of time in qualitative form (low, moderate, high, etc.) was available, whereas in some years, the information was available in the form of crop damage or losses. The entire data were converted in dichotomous form : y = 0 for low occurrence and y = 1 for medium or high occurrence. The model used was

$$P(Y=1) = \frac{1}{1+e^{-Z}} + e,$$

$$Z = a_0 + b_1X_1 + b_2X_2,$$

where $X_1$ = maximum temperature
$X_2$ = mean RH
e = random error

Prediction rule :

If $p \geq .5$,probability of occurrence of epidemic is more.
If $p < .5$, probability of occurrence of epidemic is less.

Using this model, the reliable forewarning could be obtained in May. The forewarning and the actual status for subsequent years are given in Table 14.2.

**Table 14. 2** : Forewarning of Pyrilla in Sugarcane (Muzaffarnagar)

| *Year* | *Actual status* | *Forewarning* |
|---|---|---|
| 1989 | Low | Low |
| 1990 | Low | Low |
| 1991 | Medium | Low |
| 1999 | High | High |

It can be seen that the model provided reliable forewarning except in 1991. However, in 1991 observed status was confusing because there was no mention about any pest status in any month except medium status in September. Therefore, practically, the overall status during 1991 can be assumed to be low. As such, the model is able to provide reliable forewarning (Kumar et al., 2004).

## References

Agrawal, R., Jain, R.C. and Singh, D. (1980), Forecasting of Rice Yield using Climatic Variables, *Ind. J. Agric. Sci.*, 50(9), 680-684.

Agrawal R. and Jain, R.C. (1982), Composite Model for Forecasting Rice Yield, *Ind. J . Agric. Sci.*, 52(3), 189-194.

Agrawal, R., Jain, R.C. and Jha, M.P. (1983), Joint Effects of Weather Variables on Rice Yields, *Mausam*, 34(2), 177-181.

Agrawal, R., Jain, R.C. and Jha, M.P. (1986), Models for Studying Rice Crop-Weather Relationship, *Mausam*, 37(1), 67-70.

Agrawal, R., Jain, R.C. and Mehta, S.C. (2001), Yield Forecast Based on Weather Variables and Agricultural Input on Agroclimatic Zone Basis, *Ind. J. Agric. Sci.*, 71(7), 487- 490.

Agrawal, R., Kumar, A., Bhar, L.M., Mehta, S.C., Chattopadhyay, C., Chandrashekhar, Singh, Y.P., Sharma, K.C., Meena, R.L., Roy, S.K., Singh, C.P., Ramakrishna, Y.S., Rao, S. and Prasad, Y.G. (2004), Model for Forewarning Mustard Aphid, 57th Conference of ISAS.

Jain, R.C., Agrawal R. and Jha, M.P. (1980), Effect of Climatic Variables on Rice Yield and its Forecast, *Mausam*, 31(4), 591-596.

Kumar, A., Agrawal, R., Srivastava, A.K., Srivastava, D.C., Sinha, O.K., Prasad, Y.G. and Ramakrishna, Y.S. (2004), Model for Forewarning Pyrilla in Sugarcane, 57th Conference of ISAS.

Mehta S.C., Agrawal, R. and Singh, V.P.N. (2000), Strategies for Composite Forecast, *JISAS*, 53(3), 262-272.

Mehta S.C., Agrawal, R., Shukla, R.P. and Sharma S. (2001), A Statistical Model for Prediction of Mango-fruitfly Outbreak, *National Seminar on Agrometeorological Research for Sustainable Agricultural Production.*

Rai, T. and Chandrahas (2000), Use of Discriminant Function of Weather Parameters for Developing Forecast Model of Rice Crop. *Project Report*, IASRI Publication.

Ramakrishna, Y.S. (2003), Development of Weather Based Forewarning Systems for Crop Pests and Diseases, *Project Report*.

Ramasubramanian, V., Misra, A.K. and Om Prakash (2002), Abstract : Statistical Modeling for Forewarning Epidemic Outbreak of Powdery Mildew Diseases in Mango, *JISAS*, 55(3), 336-337.

Saxena, A., Jain, R.C. and Yadav, R.L. (2001), Development of Early Warning and Yield Assessment Models for Rainfed Crops based on Agro-meteorological Indices, *Project Report*, IASRI Publication.

Trivedi, T.P., Jain, R.C., Mehta, S.C. and Bhar, L.M. (1999), Development of Forewarning System for Aphid, Myzus Persicae (Sulzer) on Potato, *Annual Report*, NCIPM, New Delhi, 43-49.

15

# Crop Forecasting using Time-series Data : An Analysis on Foodgrains in India

***Sourav Chakraborty* and *Sugata Sen Roy***

## 1. Introduction

Agriculture plays an important role in the Indian Economy. It provides gainful employment to a significant large section of Indian society and provides raw material for a large number of industries in the country's Net Domestic product. Agricultural sector is comprised of crop production, dairy, fisheries, animal husbandry, forestry etc. of which crop production is the most important activity from the point of income generation and employment potential. The agricultural activities are governed by agro-climatic conditions having divergent features spread over the length and breadth of the country.

In the whole gamut of agricultural statistics, the most dominant and significant component relate to statistics on crop production. The crop statistics revolve around two major components, viz., area and yield. The area statistics broadly cover the utilization of land with detailed statistics relating to land under agriculture in different seasons. The yield statistics deal with productivity of land in respect of various crops. The product of crop yield estimates and crop area estimates gives the production estimates. The procedure formerly used for the estimation of the yield of a crop was based on the formula

Total yield = area × normal yield x condition factor.

The traditional method of estimating the yield per hectare, called the *annawari method,* was based on the notions of the 'normal yield' and the 'condition factor'. The 'normal yield' is defined as the average yield on average soil in an average year for a particular district. The 'condition factor' referred to that of a village and showed how the yield per hectare during the given year differed from normal yield. The factor was expressed as so many annas per rupee, the rupee representing the normal yield. Because of the vagueness of the concept of 'normal yield' and also because the 'condition factor' was based on eye-appraisal by *patwaris* or *chowkidars,* the traditional method was ultimately abandoned. Currently crop-cutting experiments are conducted under a well-designed crop estimation survey for estimating the average yield in India. An estimate is obtained through actual harvesting, threshing and weighing of the crop grown in a small selected area (called 'cuts') of the selected field. A stratified multistage random sampling method is used for the selection, with tehsils (each containing 100 to 300 villages) as strata, a village as the primary unit, a field growing the particular crop as the secondary unit and a cut within the field as the ultimate sampling unit. For the system adopted for collection of area statistics, the States and Union Territories (UTs) are classified into three groups : (i) States and Union Territories where the area and land use statistics are built up through complete enumeration as a part of land records maintained by revenue agencies, (ii) Three permanently settled states of Kerala, West Bengal and Orissa where no regular agency for area enumeration exists, and a sample of villages is selected each year and covered on complete enumeration basis by statistical agencies, (iii) Some hilly districts of Assam and the North Eastern and other states which are neither cadastrally surveyed nor have the requisite revenue agency. For these areas, the statistics reported are in the nature of eye estimates of revenue officers. Out of the total reporting area, estimates for 81.7% of the area are based on complete enumeration, those for 9.2% on sample surveys and those for the remaining 9.1% on conventional or impressionistic estimates by village *chowkidars* or higher revenue officials.

The objective of this study is to develop a model by which we can estimate and forecast the production of total food grains in India. Some literatures are available regarding estimation of area, yield and production in Indian agriculture. For instance, Narayana and Parikh (1981) estimated acreage response for different crops by using expected revenue instead of expected prices as a proxy for expected profit. In this paper we seek to use the Box-Jenkins time-series models on the production, yield and area to find the forecasted values.

Generally, time-series models are used to predict the production of a particular crop on the basis of its past productions. However, since the area under the crop varies from year to year, it may be better to fit separate models to area and yield-rate so as to capture their intrinsic behaviours and then to obtain the predicted production value by multiplying the area and the yield predictions. We apply four different methods for forecasting the production of food grains as a whole and two methods for the production of each individual crop which comes under food grains. In the *first method* (*Method* 1), we model the production data for each crop and then obtain the production forecasts for it in the coming years. In the *second method* (*Method* 2), we separately model the area and yield for any crop, obtain the forecasts of area and yield, and then get the forecast values of production by multiplying the area and yield forecasts. We perform this analysis for each of the crops, Rice, Wheat, Coarse Cereals, and Pulses. Predictions for total food grains can then be made either by using Method 1 or Method 2 for total food grains or by adding the predictions of individual crops as obtained by Method 1 (Method 3) or Method 2 (Method 4).

## 2. The Model

In this study we use the Box-Jenkins model (Box et. al., 1994) for forecasting. Consider an Autoregressive-Integrated-Moving-Average process (ARIMA (p, d, q)) $x_t$,

$$\Phi(B)(1-B)^d x_t = \psi(B)e_t,$$

where B is a backward-shift operator, p and q are respectively the orders of the autoregressive and moving-average components, d is the number of times the series needs to be differenced to make it stationary and

$$\Phi(w) = 1-\alpha_1 w-\alpha_2 w^2 -\ldots- \alpha_p w^p$$

and

$$\psi(w) = 1+\beta_1 w-\beta_2 w^2 -\ldots \beta_q w^q.$$

The predictions in such models are given by the conditional mean obtained on the basis of past observations (Brockwell and Davis, 1991). However, since the parameters of a model as well as its orders are unknown, these need to be estimated first. An indication of the orders is provided by the autocorrelation and the partial autocorrelation functions. The parameters are then estimated by using the maximum likelihood method. To select a model, either Akaike's Information Criterion (AIC),

$$AIC = -2l/n + 2k/n$$

or Schwarz's Criterion,

$$SC = -2l/n + k \log n/n$$

can be used. The predictions are then made on the basis of this selected model. Labeling the years from 1950 as 1, 2, ..., we define,

$$z_t^i = \text{production of } i^{th} \text{ type of crop in } t^{th} \text{ year}$$

$$a_t^i = \text{area under } i^{th} \text{ type of crop in } t^{th} \text{ year}$$

$$y_t^i = \text{yield rate of } i^{th} \text{ type of crop in } t^{th} \text{ year}$$

for i = r (rice), w (wheat), c (coarse cereals) and p (pulses) and t = 1,2,...n (n=51).

Thus $z_t^i = a_t^i . y_t^i$ .

In Method 1, for a given *i*, we fit a model to $z_t^i$ and then predict $z_{n+s}^i$, where given n, s is the number of steps ahead that we want to predict. In Method 2, for a given i, separate models are fitted to $a_t^i$ and $y_t^i$, from which the corresponding predictions $a_{n+s}^i$ and $y_{n+s}^i$ are obtained. The production for the (n+s)[th] year is then predicted as $z_{n+s}^i = a_{n+s}^i \cdot y_{n+s}^i$. For production of food grains both Method 1 and Method 2 can be applied with i = f (total food grain). The alternatives (Method 3 and Method 4) would be to obtain the prediction as $z_{n+s}^f = \sum_{i=r,w,c,p} z_{n+s}^i$, where the $z_{n+s}^i$'s are obtained either by Method 1 or Method 2.

To compare the forecasts as obtained by different methods it is necessary to look at some measure of forecast error. The four types of forecast error statistics generally used are :

$$\text{Root Mean Squared Error (RMSE)} = \sqrt{\left\{\frac{1}{h+1}\sum_{t=s}^{s+h}(\hat{y}_t - y_t)^2\right\}}$$

$$\text{Mean Absolute Error (MAE)} = \frac{1}{h+1}\sum_{t=s}^{s+h}\left|\hat{y}_t - y_t\right|$$

$$\text{Mean Absolute Percentage Error (MAPE)} = \frac{1}{h+1}\sum_{t=s}^{s+h}\frac{\left|\hat{y}_t - y_t\right|}{y_t}$$

$$\text{Theil Inequality Coefficient (TIC)} = \frac{\sqrt{\left\{\frac{1}{h+1}\sum_{t=s}^{s+h}(\hat{y}_t - y_t)^2\right\}}}{\sqrt{\{\frac{1}{h+1}\sum_{t=s}^{s+h}\hat{y}_t^2\}} + \sqrt{\{\frac{1}{h+1}\sum_{t=s}^{s+h}\hat{y}_t^2\}}}$$

The first two measures depend on the scale of the variable. These can thus be used only as relative measures to compare forecasts for the same series across different models – the smaller the error, the better the forecasting ability of that model. The remaining two statistics are scale invariant, with the Theil inequality coefficient always lying between zero and one, where zero indicates a effect fit.

In this paper, however, we will only compute the forecast errors for the production, yield and area separately. Forecast errors when the forecast is obtained as a product of two forecast values will be considered in a future paper.

## 3. The Data

The predictions made in this paper are based on 51 years' data on production, area and yield. The source of the data is the various issues of 'Agricultural Statistics at a Glance' published by Ministry of Agriculture, Government of India each year.

From the data on production, yield and area of different crops and on total crops for the years 1950-51 to 2000-01, it can be observed that the production of rice and wheat show an upward trend, while those of cereals and pulses remain more or less constant. However, it has been observed that although the yield of pulses has remained constant over the years, cereals have had, like rice and wheat, an increasing trend. So the question arises as to why the production of cereals has not increased. This can be explained by comparing the areas for different crops, which shows, unlike the other crops, the area under cereals, after an initial increase, has actually dropped over the years. This is a major reason why we have looked at area and yield separately in predicting production in the next section.

## 4. Estimation and Forecasting of the Four Major Crops

### 4.1 Rice

The data on all three variables, production, yield and area, under rice show an increasing trend. Thus it is obvious that these series are non-stationary by nature. A choice among the different models shows that according to both the AIC and the SC, the following models give the best fit.

*Production* : ARIMA (2, 2, 1)

$$z_t^r = 1.29386 z_{t-1}^r - 0.039663 z_{t-2}^r + 0.19773 z_{t-3}^r - 0.451937 z_{t-4}^r + \varepsilon_t - 0.958597 \varepsilon_{t-1} \quad \ldots (4.1)$$

with AIC = 5.877248 and SC = 5.994198.

*Yield* : ARIMA (2, 2, 1)

$$y_t^r = 1.290419 y_{t-1}^r - 2.871632 y_{t-2}^r + 0.195359 y_{t-3}^r - 0.45247 y_{t-4}^r + \varepsilon_t - 0.952607 \varepsilon_{t-1} \quad \ldots (4.2)$$

with AIC = 11.95634 and SC = 12.07329.

*Area* : ARIMA (1,2,1)

$$a_t^r = 1.537459 a_{t-1}^r - 0.074918 a_{t-2}^r - 0.462541 a_{t-3}^r + \varepsilon_t - 0.96455 \varepsilon_{t-1} \quad \ldots (4.3)$$

with AIC =2.749626 and SC = 2.826843.

Using Method 1, the forecast value of production of rice for the four years (s =1,2,3,4) on the basis of (4.1) are as shown in Table 15.1.

**Table 15.1 :** Production Forecasts of Rice as obtained by Method 1

| *Agricultural Year* | 2002-03 | 2003-04 | 2004-05 | 2005-06 |
|---|---|---|---|---|
| *Forecast Value* | 92.86 | 92.73 | 96.29 | 97.2 |

In Method 2, on the basis of (4.2) and (4.3) we obtained the forecast values of the yield and the area for the four years separately and the forecasts for production are then obtained by multiplying the two as shown in Table 15.2.

**Table 15.2 :** Production Forecasts of Rice as obtained by Method 2

| *Agricultural Year* | 2002-03 | 2003-04 | 2004-05 | 2005-06 |
|---|---|---|---|---|
| *Yield Forecast* | 2056 | 2056 | 2132 | 2141 |
| *Area Forecast* | 45.04 | 45.22 | 45.51 | 45.75 |
| *Production Forecast* | 92.6 | 92.97 | 97.02 | 97.95 |

Comparison of the forecasts of production of rice as obtained by the two different methods as shown in Table 15.1 and 15.2 shows very little difference in the forecast values. The forecast errors shown in Table 15.3 indicate that the fit is fairly good.

**Table 15. 3 :** Forecasting Errors of Rice as measured by different methods

| | *RMSE* | *MAE* | *MAPE* | *TIC* |
|---|---|---|---|---|
| *Area* | 0.918566 | 0.702256 | 265.7009 | 0.284738 |
| *Production* | 4.293856 | 3.453770 | 227.7347 | 0.246833 |
| *Yield* | 89.72358 | 71.58134 | 68.95639 | 0.246176 |

### 4.2 Wheat

Wheat is the crop which has had a phenomenal rise in its yield over the years and although the area under the crop has risen all but slowly, production has increased about 15-folds over the second half of the twentieth century. Again choosing among the different ARIMA models, the ones which give the best fit according to the AIC and SC are :

*Production* : ARIMA (2, 2, 1)

$$z_t^w = 0.585474 z_{t-1}^w - 0.414526 z_{t-2}^w + \varepsilon_t - 0.958282 \varepsilon_{t-1} \quad \dots (4.4)$$

with AIC = 5.028824 and SC = 5.106041.

*Yield* : ARIMA (2, 2,1)

$$y_t^w = 1.456693 y_{t-1}^w - 0.291527 y_{t-2}^w + 0.212975 y_{t-3}^w - 0.378141 y_{t-4}^w + \varepsilon_t - 0.93628 \varepsilon_{t-1} \quad \dots (4.5)$$

with AIC = 12.06692 and SC = 12.18387.

*Area* : ARIMA (1,1,1)

$$a_t^w = 0.252862 a_{t-1}^w + 0.747138 a_{t-2}^w + \varepsilon_t - 1.208626 \varepsilon_{t-1} \quad \dots (4.6)$$

with AIC = 2.398770 and SC = 2.475251.

Using Method 1, the forecasted value of production of wheat for the four years (s =1,2,3,4) on the basis of (4.4) are as shown in Table 15.4.

**Table 15.4** : Production Forecasts of Wheat as obtained by Method 1

| *Agricultural Year* | 2002-03 | 2003-04 | 2004-05 | 2005-06 |
|---|---|---|---|---|
| *Forecast Value* | 72.89 | 74.4 | 75.73 | 77.13 |

In Method 2 on the basis of (4.5) and (4.6) we obtained the forecast values of the yield and the area of the four years separately forecasted and the forecasts for production are then obtained by multiplying the two as shown in Table 15.5.

**Table 15.5 :** Production Forecasts of Wheat as obtained by Method 2

| *Agricultural Year* | 2002-03 | 2003-04 | 2004-05 | 2005-06 |
|---|---|---|---|---|
| *Yield Forecast* | 2856 | 2879 | 2927 | 2985 |
| *Area Forecast* | 25.94 | 25.93 | 25.94 | 25.93 |
| *Production Forecast* | 74.08 | 74.65 | 75.93 | 77.4 |

Comparison of the forecasts of production of wheat as obtained by the two different methods as shown in Tables 15.4 and 15.5 shows a difference in the 2002-03 value but very little difference for the other years. The forecast errors shown in Table 15.6 indicate that the fit are not as good as that of rice.

**Table 15. 6 :** Forecasting Errors of Wheat as measured by different methods

| | *RMSE* | *MAE* | *MAPE* | *TIC* |
|---|---|---|---|---|
| *Area* | 0.771396 | 0.606056 | 158.6306 | 0.521004 |
| *Production* | 2.870990 | 2.271672 | 159.5741 | 0.300129 |
| *Yield* | 94.82394 | 79.09021 | 139.6072 | 0.284645 |

### 4.3 *Coarse Cereals*

Among the four crops, coarse cereals seem to be different from the other three primarily because there seems to be only a minor increase in its production over the years. However, the yield data show a significant increase though not as phenomenally as wheat. This event can be explained by the fact that the area under cereals, after an initial increase, has actually declined over the years.

Choosing through different models for the data on production, yield and area, the AIC and SC lead to the following models.

*Production* : ARIMA (1,1,1)

$$z_t^{cc} = 1.054463\, z_{t-1}^{cc} - 0.054463\, z_{t-2}^{cc} + \varepsilon_t - 0.755672\varepsilon_{t-1} \quad \ldots (4.7)$$

with AIC = 4.976850 and SC = 5.053331.

*Yield* : ARIMA (2,2,1)

$$y_t^{\alpha} = 1.431524 y_{t-1}^{\alpha} - 0.26382 y_{t-2}^{\alpha} + 0.233068 y_{t-3}^{\alpha}$$
$$-0.400772 y_{t-4}^{\alpha} + \varepsilon_t - 0.957155\varepsilon_{t-1} \quad \ldots (4.8)$$

with AIC = 11.30888 and SC = 11.42583.

*Area* : ARIMA (1,2, 1)

$$a_t^{cc} = 1.75516_{t-1}^{cc} - 0.51032 a_{t-2}^{cc} - 0.24484 a_{t-3}^{cc}$$
$$+\varepsilon_t - 0.989949\varepsilon_{t-1} \quad \ldots (4.9)$$

with AIC = 3.694980 and SC = 3.772198.

Using Method 1 the forecast value of production of coarse cereals for the four years (s=1, 2, 3, 4) on the basis of (4.7) are as shown in Table 15.7.

**Table 15.7 :** Production Forecasts of Coarse Cereals as obtained by Method 1

| *Agricultural Year* | 2002-03 | 2003-04 | 2004-05 | 2005-06 |
|---|---|---|---|---|
| *Forecast Value* | 34.73 | 35.41 | 36.08 | 36.75 |

In Method 2 the forecasts of yield and the area of the four years are done on the basis of (4.8) and (4.9) respectively and the forecasts for production are then obtained by multiplying the two as shown in Table 15.8.

**Table 15.8 :** Production Forecasts of Coarse Cereals as obtained by Method 2

| *Agricultural Year* | 2002-03 | 2003-04 | 2004-05 | 2005-06 |
|---|---|---|---|---|
| *Yield Forecast* | 1111 | 1114 | 1156 | 1162 |
| *Area Forecast* | 29.57 | 29.31 | 29.08 | 28.84 |
| *Production Forecast* | 32.85 | 32.65 | 33.62 | 33.51 |

Unlike that of rice and wheat, forecasts of production of

coarse cereals as obtained by the two different methods differ over the year with Method 2 giving smaller forecasted values. The probably is to be expected because the area under the crop decreased over the year after an initial increase. The forecast errors are shown in Table 15.9.

**Table 15.9 :** Forecast Errors of Coarse Cereals as measured by different methods

| | *RMSE* | *MAE* | *MAPE* | *TIC* |
|---|---|---|---|---|
| *Area* | 1.473640 | 1.116195 | 212.6932 | 0.327642 |
| *Production* | 2.799627 | 2.301937 | 150.5233 | 0.512047 |
| *Yield* | 64.90986 | 49.29999 | 177.0078 | 0.279843 |

### 4.4 *Pulses*

Pulses have over the years shown very little change in either its yield or its area and consequently in its production as well. Here the AIC and the SC lead to the following models :

*Production* : ARIMA (1,2,1)

$$z_t^P = 1.453202 z_{t-1}^P - 0.093596 z_{t-2}^P + 0.546798 z_{t-3}^P + \varepsilon_t - 0.973746 \varepsilon_{t-1} \quad \ldots (4.10)$$

with AIC = 3.679207 and SC = 3.756424.

*Yield*: ARIMA (1,2,1)

$$\alpha_t^{cc} = 1.75516 \alpha_{t-1}^{\alpha} - 0.51032 \alpha_{t-2}^{cc} + 0.24484 \alpha_{t-3}^{cc} + \varepsilon_t - 0.9989949 \varepsilon_{t-1} \quad \ldots (4.11)$$

with AIC = 10.748170 and SC = 10.825390.

*Area* : ARIMA (1,1,1)

$$a_t^P = 1.491373 a_{t-1}^P - 0.491373 a_{t-2}^P + \varepsilon_t - 0.989884 \varepsilon_{t-1} \quad \ldots (4.12)$$

with AIC = 2.842177 and SC = 2.918658.

Using Method 1, the forecast value of production of pulses for the four years (s = 1,2,3,4) on the basis of (4.10) are as shown in Table 15.10.

**Table 15.10 :** Production Forecasts of Pulses as obtained by Method 1

| *Agricultural Year* | 2002-03 | 2003-04 | 2004-05 | 2005-06 |
|---|---|---|---|---|
| *Forecast Value* | 12.18 | 12.87 | 12.63 | 12.89 |

In Method 2, the yield and the area of the four years are separately forecasted on the basis of (4.11) and (4.12) respectively and the forecasts for production are then obtained by multiplying the two as shown in Table 15. 11.

**Table 15. 11 :** Production Forecasts of Pulses as obtained by Method 2

| *Agricultural Year* | 2002-03 | 2003-04 | 2004-05 | 2005-06 |
|---|---|---|---|---|
| *Yield Forecast* | 578 | 602 | 594 | 604 |
| *Area Forecast* | 22.25 | 22.54 | 22.68 | 22.75 |
| *Production Forecast* | 12.68 | 13.57 | 13.47 | 13.74 |

Forecasts of production of pulses show very little in change, particularly for Method 1. The forecast as obtained by Method 2 gives very small increase s over the years. The forecast errors are shown in Table 15.12.

**Table 15.12 :** Forecast Errors of Pulses as measured by different methods

| | *RMSE* | *MAE* | *MAPE* | *TIC* |
|---|---|---|---|---|
| *Area* | 0.962858 | 0.793214 | 159.5720 | 0.536527 |
| *Production* | 1.462063 | 1.134305 | 90.81315 | 0.266798 |
| *Yield* | 50.11544 | 37.98419 | 113.4690 | 0.254906 |

## 5. Forecasting Food Grains

As stated before, the prediction of food grains can be obtained by four different methods. It can be obtained by applying Methods 1 and 2 on the food grain figures or just by summing the predictions of the four crops as obtained through Methods

1 and 2 of Section 4. We will henceforth call these methods as Methods 3 and 4, respectively.

Using the two model selection criteria, the AIC and the SC, the chosen models for production, yield and area are :

*Production* : ARIMA (2,2,1)

$$z_t^f = 0.31641 z_{t-1}^f - 0.347628 z_{t-2}^f + 0.335962 z_{t-3}^f + \varepsilon_t - 0.980983 \varepsilon_{t-1} \quad \ldots (5.1)$$

with AIC = 7.349773 and SC = 7.466723.

*Yield* : ARIMA (2,2,1)

$$y_t^f = 1.407192 y_{t-1}^f - 0.164139 y_{t-2}^f + 0.106702 y_{t-3}^f - 0.349755 y_{t-4}^f + \varepsilon_t - 0.932075 \varepsilon_{t-1} \quad \ldots (5.2)$$

with AIC = 11.08474 and SC = 11.20169.

*Area* : ARIMA (1,1,0)

$$a_t^f = 0.401273 a_{t-1}^f + 0.598727 a_{t-2}^f \quad \ldots (5.3)$$

with AIC = 5.003459 and SC = 5.045681.

Using Method 1, the forecast values of production of pulses for the four years (s =1,2,3,4) on the basis of (5.1) are as shown in Table 15.13.

**Table 15.13 :** Production Forecasts of Food Grains as obtained by Method 1

| *Agricultural Year* | 2002-03 | 2003-04 | 2004-05 | 2005-06 |
|---|---|---|---|---|
| *Forecasted Value* | 212.66 | 213.79 | 219.48 | 221.88 |

In Method 2, the forecast values of yield and the area of the four years are separately forecasted on the basis of (5.2) and (5.3), respectively and the forecasts for production are then obtained by multiplying the two as shown in Table 15.14.

**Table 15.14 :** Production Forecasts of Food grains as obtained by Method 2

| *Agricultural Year* | 2002-03 | 2003-04 | 2004-05 | 2005-06 |
|---|---|---|---|---|
| *Yield Forecast* | 1757 | 1765 | 1812 | 1839 |
| *Area Forecast* | 121.39 | 121.69 | 121.50 | 121.61 |
| *Production Forecast* | 213.28 | 214.78 | 220.16 | 223.64 |

The forecasts as obtained by Method 3 and Method 4 are shown in Table 15.15.

**Table 15 :15 :** Production Forecasts of Food grains as obtained by Methods 3 and 4.

| *Agricultural Year* | 2002-03 | 2003-04 | 2004-05 | 2005-06 |
|---|---|---|---|---|
| Production Forecast Method 3 | 212.66 | 215.41 | 220.73 | 223.97 |
| Production Forecast Method 4 | 212.39 | 213.84 | 220.04 | 222.60 |

Table 15.16 shows the forecast errors of Foodgrains as measured by different methods.

**Table 15.16 :** Forecast Errors of Food grains as measured by different methods

| | *RMSE* | *MAE* | *MAPE* | *TIC* |
|---|---|---|---|---|
| *Area* | 2.856161 | 2.242519 | 132.0643 | 0.512144 |
| *Production* | 8.966076 | 7.312292 | 94.99768 | 0.255634 |
| *Yield* | 58.02844 | 49.23265 | 210.9601 | 0.277430 |

As can be observed from Tables 15.13, and 15.14 and 15.15, the forecast values do not differ by much for any of the four predicted years.

## 6. Concluding Remarks

For individual crops, the forecasts obtained by Method 2 are on the higher side than that obtained by Method 1 except in the case of coarse cereals where the situation is reversed. The reversed nature in coarse cereals is expected because there is a decline in area under coarse cereals over the years under study. Method 2 in turn captures this declining trend and thus the forecasted production slightly goes down. Overally, it seems that Method 2 gives a better prediction than the Method 1 for individual crops.

Among the four different methods used for forecasting the production of food grains, intuitively there exists a similarity between Methods 1 and 3 and between Methods 2 and 4. From Tables 15.13-15.16, it is clear that Method 3 gives a higher forecast than Method 1. But between Methods 2 and 4, Method 2 gives higher predictions.

The results as obtained by the four different methods need to be compared in terms of the forecasting errors. However, for Methods 2 and 4, where the production forecasts are obtained through a multiplication of yield and area forecasts, the measures are not so easy to calculate. This is a problem that we hope to address in our future work. As for the area under crops, the total area shows an initial upward trend and then stabilizes approximately around the agricultural year 1959-60 onwards. The year 1959 is important in the context of the Indian Economy. The First Five Year Plan in India covered a five-year period from April 1951 to March 1956. Agriculture obtained the highest priority in the First Plan.

According to the Review of the First Plan, of the total plan provision of Rs.2,378 crores for the public sector, agriculture and community development received Rs. 354 crores (14.9 per cent), irrigation and power received Rs. 647 crores (27.2 per cent). Together these two sectors received Rs.1001 crores (42.09 per cent) out of the outlay of Rs.2,378 crores. The First Plan's emphasis on agriculture was fully justified because in the pre-planning period there were major food shortages in the country. Agriculture production had to be stepped up with the purpose of making India more than self-sufficient in respect of food and raw material. As a result of all these efforts, agriculture production recorded an increase of 18 per cent during 1951-56. Food production rose from 50 million tonnes before the Plan to about 65 million tonnes in the last year of the plan. From irrigation schemes, included in the First Five Year Plan, an area of about 8.5 million acres was irrigated by 1955-56. Thus it is expected that the effect of these initial boost to the agricultural sector will increase the land use substantially. However, although the total area has remained more or less constant since the early 1660s, there are major shifts in the area under different crops with wheat gaining the most and the area under coarse cereals actually declining.

## References

Box, G.E.P., Jenkins, G.M. and Reinsel, G.C. (1994), *Time Series Analysis : Forecasting and Control,* Prentice Hall, Englewood Cliffs, New Jersey.

Brockwell, P.J. and Davis, R.A. (1991), *Time Series : Theory and Methods,* Springer-Verlag, New York.

Narayana, N.S.S. and Parikh, K.S. (1981), Estimation of Farm Supply Response and Acreage Allocation : A Case Study of Indian Agriculture, Presented at the International Institute for Applied Systems Analysis, Laxemburg, Austria.

Publication of Ministry of Agriculture, Government of India (2000), *Agricultural Statistics at a Glance.*

# 16

# Statistics for Impact Analysis

*Debasis Bhattacharya*

## 1. Introduction

When we conduct any experiment related to some sociological issues, agricultural consumer behaviour problems or of any other types, we are eager to know the result of it. Moreover, when we bring some changes to the existing method (say, teaching) or introduce a new method by discarding an old method we are interested to see how the changes work. Doctors experimenting with a new therapy want to know the effectiveness of the therapy to a group of patients. An agricultural scientist might be interested to see the effectiveness of some training programme, i.e., to see whether the training programme is just right or needs modification, or he may be interested to see the impact of adoption of some agricultural practices on the yield etc. In all the cases we have to develop some effectiveness measure which the area of impact analysis encompasses. Impact Analysis is a domain where Statisticians can play around freely with the available tools. What a statistician requires for impact analysis are the data before and data after the programme for which the impact analysis is sought. Usable tools are different for different kinds of programmes and data. Data are of two types : quantitative and qualitative. Before getting the impact analysis procedure started three most important points are to be noted :

(1) Nature of the data
(2) Conditions under which data are obtained
(3) Appropriate tools for impact analysis to be used for available the data.

We can have the flow chart (Fig. 16.1) for any impact analysis procedure :

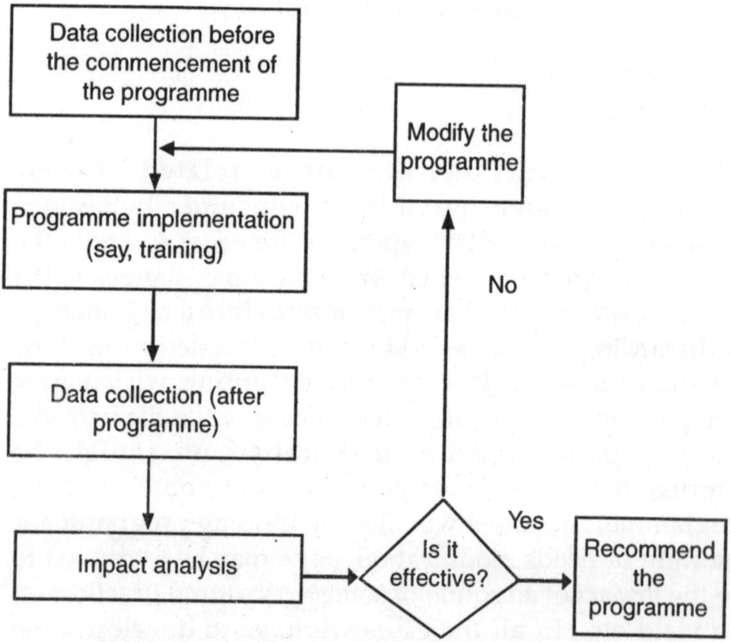


**Fig 16.1** : Flow chart for any impact analysis procedure

It is to be noted that for some programme the data after the programme are immediate and for some programme we have to observe the units after a considerable time gap. We will now discuss how the impact analysis procedures work under the following queries :

(1) What if we have quantitative data?
(2) What if we have qualitative data?

## 2. Quantitative Data

Here simple descriptive measures like, mean or median, standard deviation or quartile deviation, coefficient of variation or coefficient of quartile deviation etc. can be used for the purpose of analysis. Moreover, various parametric test procedures are available to see the significant change in the level of mean, dispersion, difference in proportions etc. Suppose, we want to evaluate the impact of some programme or experiment in changing (increasing, say) the mean level of response from $\mu_0$ to $\mu$, where $\mu > \mu_0$, $\mu_0$ is the hypothesized mean of response set from our previous experience, and $\mu$ is the mean response in the population obtained after conducting the experiment or programme. Here we set our null hypothesis as $H_0$: $\mu = \mu_0$ and alternative hypothesis as $H_1 : \mu > \mu_0$. Acceptance of $H_1$ will imply that the programme or experiment has significant positive impact in increasing the hypothetical mean response. Assuming the population of responses being normal and depending on the situation that population variability $\sigma^2$ is known or unknown we can use very standard statistical test procedures of *Z*-test and Student's *t*-test, respectively. Any standard text of statistics discusses all these tests in detail and we here prefer to omit the discussion on how to carry out the test. Rao (1971), Lehmann (1998) may be referred to in this context.

Sometimes, there are two independent groups of sample units drawn from two independent populations of same type. Considering one group as the Programme Group (*PG*) on which the programme or the experiment has been performed and the other group as Non-Programme Group (*NPG*) or control group. We may be interested to measure the impact of the programme or experiment. Let $\mu_{PG}$ be the mean of the programme group and $\mu_{NPG}$ be the mean of the non-programme group. Then we set our hypotheses as $H_0 : \mu_{PG} = \mu_{NPG}$ and $H_1$: $\mu_{PG} > \mu_{NPG}$, and the programme or experiment has significant positive impact when $H_1$ is accepted. For example, suppose we want to evaluate the impact of a new teaching method. What we can do is we can divide the whole class in two groups. The first group is trained or taught under the new teaching method for a considerable time period and the second is not. The scores of the students belonging to the two groups for a certain examination are recorded. Here the new method will have significant positive impact if the average score for the first group is significantly higher than the average score of the second group. Here under the assumption of normality of scores standard *t*-test procedures are available. We are describing the tests in brief.

Let $X_1, X_2, \ldots, X_m$ be the observations on *m* sample units of programme group and $Y_1, Y_2, \ldots, Y_n$ be the observations on *n* sample units belonging to non- programme group.

We set our hypotheses as $H_0: \mu_{PG} = \mu_{NPG}$ and $H_1: \mu_{PG} > \mu_{NPG}$.

We first check whether $\sigma^2_{PG} = \sigma^2_{NPG}$ by carrying out an F-test where $\mu_{PG}$ and $\mu_{NPG}$ are population means for the programme group and non-programme group, respectively, and $\sigma^2_{PG}$ and $\sigma^2_{NPG}$ are the population variances for the programme group and non-programme group, respectively.

We wish to test $H_0: \sigma^2_{PG} = \sigma^2_{NPG}$ against $H_1: \sigma^2_{PG} \neq \sigma^2_{NPG}$.

We use the test statistic

$$F = \frac{S^2_{PG}}{S^2_{NPG}} \sim F_{m-1,n-1},$$

where $S^2_{PG} = \frac{1}{m-1}\sum_{i=1}^{m}(X_i - \bar{X})^2$ and $S^2_{NPG} = \frac{1}{n-1}\sum_{i=1}^{n}(Y_i - \bar{Y})^2$.

Reject $H_0$ at $100\alpha$ % level if $F_{observed} > F_{\alpha/2,\, m-1,\, n-1}$ or

$$F_{observed} < F_{1-\frac{\alpha}{2},\, m-1,\, n-1}.$$

Now we have the following two situations :

(i) If $H_0 : \sigma^2_{PG} = \sigma^2_{NPG}$ is not rejected, then we follow Fisher's *t*-test, as described below, for testing $H_0 : \mu_{PG} = \mu_{NPG}$ against $H_1 : \mu_{PG} > \mu_{NPG}$.

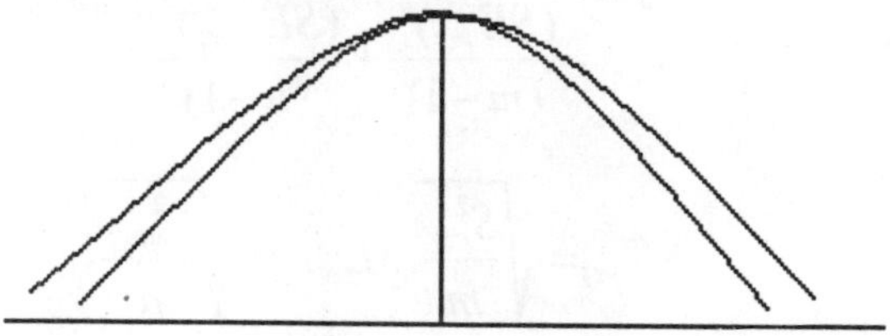

$\mu_{PG} = \mu_{NPG}$

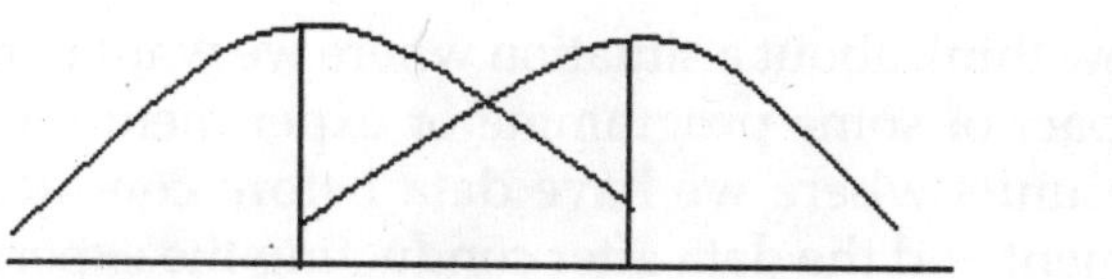

$\mu_{NPG}$ $\mu_{PG}$

Appropriate test statistics is $t = \dfrac{\bar{x} - \bar{y}}{S_p \sqrt{\{(1/m) + (1/n)\}}} \sim t_{m+n-2}$ (under $H_0$),

$$\text{where } S^2_p = \frac{(m\text{-}1)\, S^2_{PG} + (n\text{-}1) S^2_{NPG}}{(m+n\text{-}2)}.$$

Reject $H_0$ if $t_{observed} > t_{\alpha, m+n-2}$ or, alternatively, one may calculate *p*-value and reject $H_0$ if *p*-value is less than $\alpha$, where *p*-value $= P\,(t_{m+n-2} \geq t_{observed})$.

(ii) If $H_0 : \sigma^2_{PG} = \sigma^2_{NPG}$ is rejected, then we follow Welch's *t*-test, as described below, for testing $H_0 : \mu_{PG} = \mu_{NPG}$ against $H_1 : \mu_{PG} > \mu_{NPG}$.

Test statistic (under $H_0$) is $t = \dfrac{\bar{x} - \bar{y}}{\sqrt{\dfrac{S_{PG}^2}{m} + \dfrac{S_{NPG}^2}{n}}} \sim t_{d.f.}$,

where
$$d.f. = \frac{(SE_{PG})^2 + (SE_{NPG})^2}{\dfrac{(SE_{PG})^4}{(m-1)} + \dfrac{(SE_{NPG})^4}{(n-1)}},$$

$$SE_{PG} = \sqrt{\frac{S_{PG}^2}{m}}, SE_{NPG} = \sqrt{\frac{S_{NPG}^2}{n}}.$$

Reject $H_0$ at significance level $\alpha$ if $t_{observed} > t_{\alpha, d.f.}$.

Now think about a situation where we want to measure the impact of some programme or experiment on a set of sample units where we have data before conducting the experiment and the data after conducting the experiment.

Now suppose we want to measure the impact of the programme or experiment in a situation where we have data or observations before conducting the programme or experiment and the data after the end of the programme. We set our hypothesis $H_0 : \mu_A = \mu_B$ and $H_1 : \mu_A > \mu_B$ ($A$ : after the experiment, $B$ : before the experiment). Since the observations are taken on the same set of sample units we have a dependent situation. For example, we might be interested to see the impact of some weight-gaining diet programme on the basis of observations on weight before and after commencement of the programme on the same set of sample units. There, to measure the significant impact of the programme, we can use a paired $t$-test. Let $X_1, X_2, \ldots, X_n$ be the observations obtained after the experiment and $Y_1, Y_2, \ldots, Y_n$ be the observations obtained before the experiment.

Here we find $d_i = X_i - Y_i$, $i = 1, 2 \ldots, n$, and obtain $\bar{d} = (1/n)\sum_{i-1}^{n} d_i$

and $s_d^2 = \dfrac{1}{(n-1)} \sum_{i-1}^{n} (d_i - \bar{d})^2$ .

Appropriate test statistic is $t = \dfrac{\bar{d}}{s_d / \sqrt{n}} \sim t_{n-1}$ .

We reject $H_0$ at level $\alpha$ if $t_{observed} > t_{\alpha, n-1}$.

Alternatively, we compute $p$-value = $P(t_{n-1} > t_{observed})$ and reject $H_0$ if $p$-value $<\alpha$.

## 3. Qualitative Data

The procedures described above fail when we have qualitative data. There the procedures of measuring impact of some programme depends on how we can quantifies the data under proper scaling system. We might be interested to study the impact of training on knowledge or skill or we want to evaluate the extent of adoption of participants who practise for two different time points which are far apart.

In the first case respondents were asked some questions, say 10, on few points, to judge their existing knowledge.

Next we evaluate the knowledge scores before and after.

To evaluate the impact of some training programme on knowledge we have to obtain the knowledge score before and after implementation of the programme. Here before the programme the respondents are asked on few relevant questions, say 10, to judge their existing knowledge.

Let $X_i$ be the knowledge score of the $i^{th}$ participant,

$X_i = \sum_{j=1}^{n} u_{ij}, i = 1,2,\ldots,m,$ where $n$ = number of questions asked to all respondents.

$u_{ij}$ = 1, if $i^{th}$ respondent answers the $j^{th}$ question correctly
= 0, otherwise, i.e., if the answer to the $j^{th}$ question is wrong.

The importance of having knowledge on different questions might be different. There we can assign different weights to different question according to their relative importance. Let $w_j$ be the weight assigned to the $j^{th}$ question. Then weighted knowledge score is

$$X_i = \sum_{j=1}^{n} w_j u_{ij}, i = 1,2,\ldots,m.$$

For example, if the training programme is on some agricultural practice we may consider the following points on which some questions may be asked to evaluate the impact of the programme :

(1) different varieties (2) seed rates (3) selection of crops to be grown (4) seed treatment chemicals (5) inter-cropping (6) spacing (7) fertilizer dose (8) insect and plant protection chemicals (9) diseases (10) post-harvest technology.

We want to see whether the average (or location) of the score distribution shifts or not. We set our hypotheses as $H_0{:}\eta_A{=}\eta_B$ and $H_1{:}\eta_A{>}\eta_B$ ($A$ : after the programme, $B$ : before the programme).
We wish to see whether the average (or location) of the score distribution shifts. For this we test

$$H_0 : \eta_A = \eta_B \text{ vs. } H_1 : \eta_A > \eta_B .$$

| Sl. No. | Knowledge score before training | Knowledge score after training |
|---|---|---|
| 1 | 1 | 3 |
| 2 | 1 | 2 |
| 3 | 0 | 3 |
| . | . | . |
| . | . | . |
| m | 1 | 3 |

A non-parametric test will be suitable to verify $H_0$.

Wilcoxon signed ranks test for paired data will be appropriate here :

We compute $d_i = X_i - Y_i$; some are positive and some are negative.

Rank $|d_i|$'s.

Put the sign of $d_i$ 's to its rank.

$T^+$ = sum of positive ranks
$T^-$ = sum of negative ranks = $m(m+1)/2 - T^+$.

We test

$H^*_0 : \eta_A - \eta_B = 0$ against $H^*_1 : \eta_A - \eta_B > 0$.

Here $p$-value = $P\,(T^+ \geq T^+_{observed})$, which can be obtained from the Table for Wilcoxon signed ranks test.

If $p$-value $< \alpha$ (= level), then reject $H_0$, i.e., median after training has shifted to the right of the median before training, which implies that there is a significant impact of training on increasing knowledge.

If the number of participants is very large ($m \geq 15$, in practical situations '$m$' is usually more than 15), then we can use large sample test statistic. We find

$$z = \frac{T^+ - \mu_T^+}{\sigma_{T+}} \sim N\,(0,1),$$

where $\mu_T^+ = \dfrac{m(m+1)}{2}, \sigma_T^2 = \sqrt{\dfrac{m(m+1)(2m+1)}{24}}$ .

Reject $H_0$ if $z_{observed} > z_{\alpha}$, where $z_{\alpha}$ is the upper $100\alpha$ % point of a standard normal distribution, which can be obtained from the table of cumulative density for standard normal variate.

If tie occurs, we obtain the value of test statistic

$$Z = \frac{T^{+} - \mu_T^{+}}{\sqrt{\sigma_T^{+} - (1/2)\sum_{j=i}^{g} t_j(t_j - 1)(t_j + 1)}},$$

where there are $g$ ties, and $t_j$ is the length of the $j^{th}$ tie.

For example, when we have data like 1, 2, 4, 4, 4, 6, 7, 8, 9, 9, we say that there exist two ties (g=2) of length $t_1 = 3$ and $t_2 = 2$.

## 3. Conclusion

Here we have identified and explained some basic statistical methods which can be effectively used for impact analysis purposes. The methods described here are easy to apply and interpret. Some sophisticated statistical methods are also available for this purpose, especially when the data involved exhibit some special features like missing data, presence of direct or indirect effects of other covariates etc., which can be a topic of further discussion.

## References

Lehmann, E. (1998), *Non-parametrics : Statistical Methods Based on Ranks*, Pearson, New York.

Rao, C.R.( 1971), *Advanced Statistical Methods in Biometric Research*, Heffner, New York.

# 17

# A Study on Probability Distribution of Interarrival Time for Dry and Wet Years of Monsoon Rainfall

***Banjul Bhattacharyya and Swaraj Kumar Mukhopadhyay***

## 1. Introduction

An important feature of the climatic conditions is dependent upon the monsoon rainfall of a region, which does not remain constant over the years; it may exceed or fall short of the average causing flood or drought in extreme conditions. Due to this randomness of monsoon phenomenon, a prior knowledge about the pattern of rainfall is of great attraction for planning and management of water resource system as well as agricultural production process. The interarrival time between the occurrences of successive dry and wet years are of great interest to study in Sub-Himalayan region of West Bengal.

Pioneering work to study rainfall deficiency in India was done by Walker (1919). After that hardly any attempt was made in this field. Again 1960s onwards, the technologists started to work on precipitation phenomenon. To determine the annual rainfall anomalies an index was made to determine climatic fluctuations by several authors. The work of Barndroff and Yeo (1969) is noteworthy in this field. Chowdhury et al. (1989) studied the variability in drought indices with a statistical approach. But they considered the probability distribution of the interarrival time following Poisson distribution with fixed interval of time. Mukhopadhyay (1992) started to work on the field with a view that the mean value of the probability distribution is a function of time.

The aim of the paper is to determine the distribution function of interarrival time between the events in a given period of time and to test the homogeneity of the sets of same event but at different places. Also the paper targets to determine the return period of the wet years utilizing the frequency distribution.

## 2. Data and Location

In this study, 102 years (1901 - 2002) monsoon rainfall data (from June to September) have been used at two meteorological observatory centres *viz.* Jalpaiguri ($26^0 20'$ N and $89^0 27'$ E) and Malda ($25^0 02'$ N and $88^0 08'$ E) of sub-Himalayan meteorological region of India. About 80% of the annual rainfall occurs during this period.

## 3. Method

With a view to identifying the wet and dry years of monsoon rainfall at both the places, a suitable index termed as "Rainfall Index" is developed in this study. Following the criterion suggested by Mooley and Parthasarathi (1983), the rainfall index is obtained as :

$$RI = \left[\frac{(x_i - \mu)}{\sigma}\right],$$

where $x_i$ is the rainfall of $i^{th}$ years, $\mu$ is the normal rainfall and $\sigma$ is the standard deviation of the respective station. The index possesses a dimensionless number with positive and negative values. The positive value of the index indicates the wet years whereas the negative value stands for the dry years. Historical rainfall time series have been used to determine the sequences of dry and wet years separately for the two stations. The time interval between the successive dry and wet years has been measured in this study to determine the distribution of successive dry as well as wet years interarrival structure.

In general, the probability distribution of the interarrival time follows Poisson distribution with fixed interval of time, which is given by

$$p\{X = x|\mu\} = \frac{\mu^x}{x!}e^{-\mu}, x = 0,1,2,...$$

In the present study $\mu$ is not a constant but is a random variable or a random function of time 't'. A Poisson process whose intensity function $\mu(t)$ is a stochastic process is called a conditional Poisson process. It is often observed the parameter $\mu$ has the following probability density function called the gamma distribution (Ochi, 1989 and Medhi, 1996)

$$f(\mu) = \frac{1}{\Gamma(m)}\lambda^m \mu^{m-1} e^{-\lambda\mu}, \ 0 \leq \mu \leq \infty,$$

where $m$, $\lambda$ are shape and scale parameters, respectively, of the gamma distribution and $\Gamma(m)$ is the gamma function of $m$. Then, the unconditional probability distribution $P\{X = x\}$, can be obtained as

$$P\{X=x\}=\int_0^\infty \frac{\mu^x}{x!}e^{-\mu}.\frac{1}{\Gamma(m)}.\lambda^m\mu^{m-1}e^{-\lambda\mu}d\mu$$

$$=\frac{\lambda^m}{\Gamma(m)}\int_0^\infty \frac{\mu^{x+m-1}e^{-\mu(\lambda+1)}}{x!}.d\mu$$

$$=\frac{\lambda^m}{\Gamma(m)}.\frac{\Gamma(x+m)}{x!(\lambda+1)^{x+m}}$$

$$=\frac{(x-m+1)!}{(m-1)!x!}\left(\frac{1}{\lambda+1}\right)^x\left(\frac{\lambda}{\lambda+1}\right)^m.$$

By putting $\frac{1}{\lambda+1}=p$ and $\frac{\lambda}{\lambda+1}=q$, the above relation takes the form

$$P\{X=x\}=\binom{x+m-1}{x}p^xq^m$$

$$=(-1)^x\binom{-m}{x}p^xq^m,\ x=0,1,2,.....,$$

which is the negative binomial probability distribution. Thus it has been shown that the Poisson distribution with parameter $\mu$, which obeys the gamma probability law is a negative binomial distribution. [Kendall et al. (1983)]. In this distribution of time interval problem $m$ is necessarily a positive number, which is exactly $x$ failures preceding the $m^{th}$ success.

Method of moments has been applied to estimate the means and variance of the four sequences. These values are used to estimate the value of $m$. These are estimated in both wet and dry years for both the places. Data set of the individual centres has been combined together to produce larger regional data samples. The combined sequences in the arrival time of any dry and wet years can be analyzed similar to the procedures adopted for individual sequence.

The chi-square test is applied to test the goodness of fit under the null hypothesis of good agreement. To test the homogeneity of two sets of wet and dry sequences, the chi-square distribution is applied with suitable degree of freedom. Using the recurrence relation expected values of the distributions has been estimated for each of distribution separately.

## 4. Return Period

Return period of run of wet years of a certain length is calculated as the ratio of the number of recorded years to the rank number of the specified run following the method as suggested by Singh (1973).

All recorded runs of wet years obtained from seasonal rainfall data of several years are first arranged in descending order according to their length and then given rank numbers. If the run of a particular length has occurred 'n' times and there are 'r' runs of length higher than this run, then the rank number of this particular run is taken as $r + n$. The return period of run of any length is then calculated as $T/R$, where T is the number of years of record and $R = (r + n)$, is the rank number.

## 5. Results and Discussions

After preparing the rainfall index, the positive and negative values of the indices are presented in Figs. 17.1 and 17.2 and the peaks are indicated with a view to identifying the wet sequences whereas the troughs show the dry sequences for both the places. Six sets, two for wet sequences, two for dry sequences of the observatory centres and two (one dry and one wet) for the combined data are presented in Tables 17.1, 17.2 and 17.3, respectively.

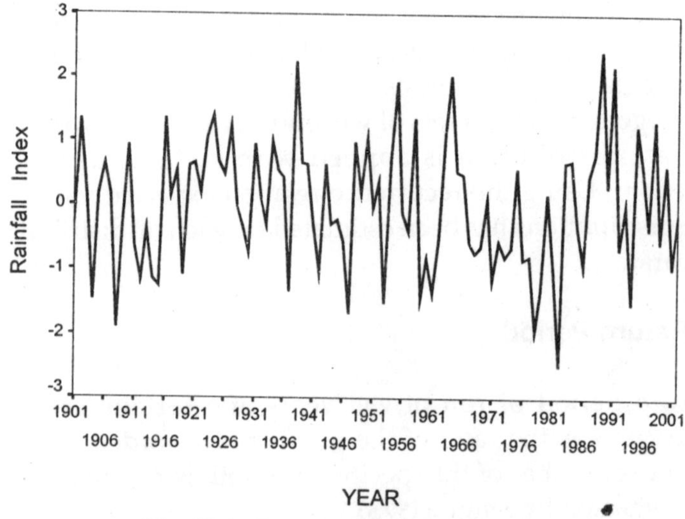


**Fig. 17.1 :** Rainfall Index for Jalpaiguri

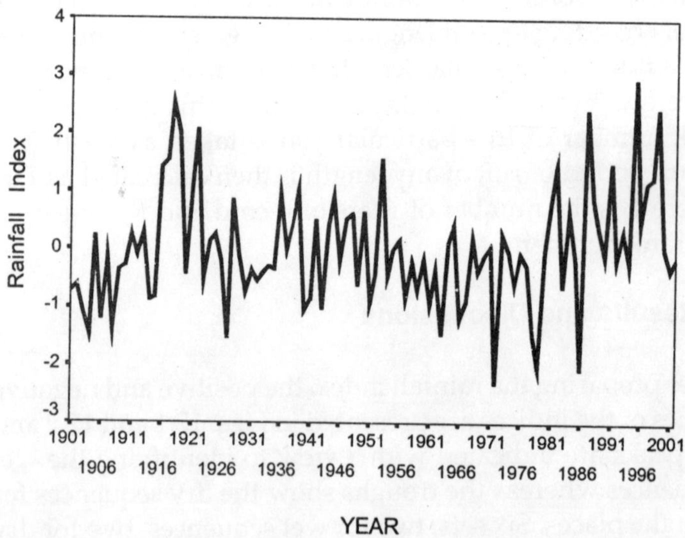


**Fig. 17.2 :** Rainfall Index for Malda

**Table 17.1 :** Observed and expected values of dry and wet sequences for Jalpaiguri

| *Wet years* | | *Dry years* | |
|---|---|---|---|
| *Observed* | *Expected* | *Observed* | *Expected* |
| 29 | 26.76 | 26 | 25.23 |
| 10 | 13.22 | 10 | 11.16 |
| 6 | 6.34 | 5 | 5.72 |
| 3 | 3.01 | 5 | 3.07 |
| 3 | 1.42 | 1 | 1.68 |
| 1 | 1.26 | 1 | 0.93 |
| | | 0 | 0.52 |
| | | 0 | 0.29 |
| | | 1 | 0.16 |
| $\chi^2_{cal} = 0.73^{NS}$ | df = 1 | $\chi^2_{cal} = 0.937^{NS}$ | df = 1 |

**Table 17. 2 :** Observed and expected values of dry and wet sequences for Malda

| *Wet years* | | *Dry years* | |
|---|---|---|---|
| *Observed* | *Expected* | *Observed* | *Expected* |
| 20 | 21.86 | 31 | 31.26 |
| 13 | 10.08 | 14 | 13.64 |
| 7 | 5.46 | 7 | 6.09 |
| 1 | 3.11 | 1 | 2.74 |
| 1 | 1.81 | 2 | 1.24 |
| 0 | 1.07 | 0 | 0.56 |
| 0 | 0.64 | 1 | 0.47 |
| 3 | 0.97 | | |
| $\chi^2_{cal} = 0.509^{NS}$ | df = 1 | $\chi^2_{cal} = 0.95^{NS}$ | df = 1 |

**Table 17. 3 :** Observed and expected values of dry and wet sequences for combined distribution

| *Wet years* | | *Dry years* | |
|---|---|---|---|
| *Observed* | *Expected* | *Observed* | *Expected* |
| 49 | 49.18 | 57 | 56.85 |
| 23 | 22.64 | 24 | 24.33 |
| 13 | 11.62 | 12 | 11.74 |
| 4 | 6.17 | 6 | 5.87 |
| 4 | 3.33 | 3 | 2.99 |
| 1 | 1.82 | 1 | 1.54 |
| 0 | 0.98 | 1 | 0.80 |
| 3 | 1.23 | 0 | 0.42 |
| | | 1 | 0.46 |
| $\chi^2_{cal} = 0.92^{NS}$ | df = 1 | $\chi^2_{cal} = 0.98^{NS}$ | df = 1 |

By the method of moments the mean, variance and the value of *m* as derived from these statistics, are calculated and presented in Table 17.4. The expected values of the probability distribution as estimated are presented in Tables 17.1, 17. 2 and 17. 3 along with the observed values. The chi-square values are also given in the same table with appropriate degrees of freedom. The frequency analysis for time interval between successive wet and dry years is best fitted to negative binomial distribution in both the places. All computed chi-square values are found to be non-significant, suggesting that there are no significant difference between the observed interarrival time of each weather event and the expected value of the fitted distribution in this region.

The chi-square values as obtained for combined dry and wet years of the two stations indicate that the two negative binomial distributions for each event are identical and their rainfall populations of the respective events are of the same type, under null hypothesis. The fits have been found to be good for each sequence.

**Table 17. 4 :** Basic statistics of negative binomial distribution for different sequences

| *Location* | *Sequence* | *Mean* | *Variance* | *m* |
|---|---|---|---|---|
| Jalpaiguri | Wet | 0.923 | 1.725 | 1.063 |
| | Dry | 1.061 | 2.547 | 0.758 |
| Malda | Wet | 1.222 | 3.24 | 0.741 |
| | Dry | 0.804 | 1.479 | 0.956 |
| Combined | Wet | 1.061 | 2.45 | 0.812 |
| | Dry | 0.924 | 1.994 | 0.797 |

Return period of run of wet years as calculated are presented in Table 17.5 and the values of runs are plotted against it are shown in Figures 17.3 and 17.4. It is found that the curves so fitted are of the form of straight lines. The graph suggests that for Malda district in 7, 10, 13 and 18 years' values of length of run of wet years are 2, 3, 4 and 5 years, respectively. For Jalpaiguri district the same length of run of wet years may occur in 11, 16, 21 and 29 years, respectively.

**Table 17. 5** : Estimated return periods of wet years

| *Location* | *Maximum length of wet years likely to occur once in T years* | | | |
|---|---|---|---|---|
| | *T* 2 | *3* | *4* | *5* |
| Jalpaiguri | 11.4 | 15.8 | 20.6 | 29.4 |
| Malda | 7.03 | 9.7 | 12.7 | 18.5 |

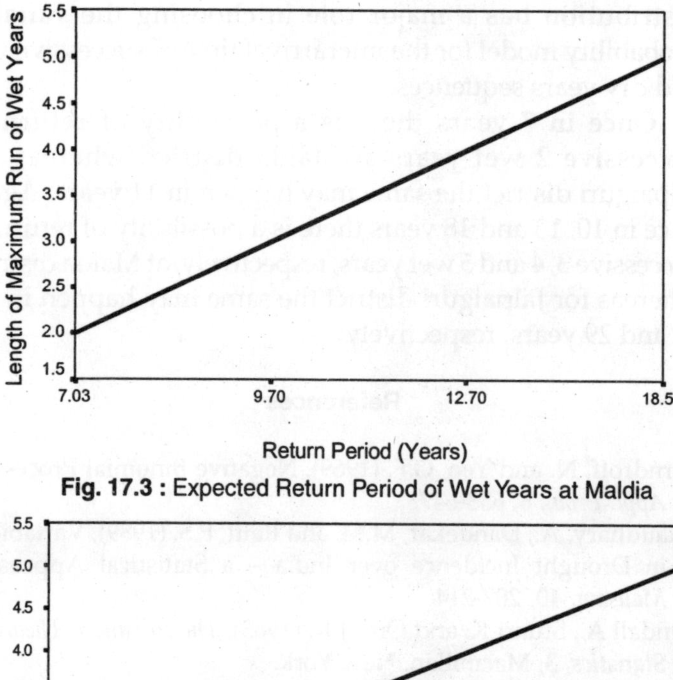


**Fig. 17.3** : Expected Return Period of Wet Years at Maldia

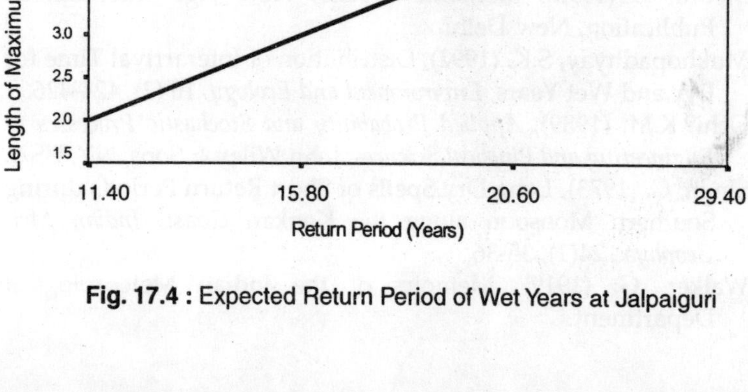


**Fig. 17.4** : Expected Return Period of Wet Years at Jalpaiguri

## 6. Conclusion

The interarrival time of dry and wet years as well as the combined sequences of same event is best fitted with method of moments of negative binomial distribution. The combined sequences of interarrival time for dry and wet years are also homogeneous in nature.

The frequency distribution of mean value of Poisson distribution has a major role in choosing the suitable probability model for the interarrival time of successive wet and dry years sequences.

Once in 7 years there is a possibility of return of successive 2 wet years at Malda district, whereas for Jalpaiguri district the same may happen in 11 years. Again once in 10, 13 and 18 years there is a possibility of return of successive 3, 4 and 5 wet years, respectively, at Malda district, whereas for Jalpaiguri district the same may happen in 16, 21 and 29 years, respectively.

## References

Barndroff, N. and Yeo, G.F. (1969), Negative Binomial Process, *J. Appl. Prob.*, 6, 633-647.

Chaudhary, A., Dandekar, M.M. and Raut, P.S. (1989), Variability in Drought Incidence over India – a Statistical Approach, *Mausam*, 40, 207-214.

Kendall A., Stuart K. and Ord, J.K. (1983), *The Advanced Theory of Statistics*, 3, Macmillan, New York.

Medhi, J. (1996), *Stochastic Process*, New Age International Publication, New Delhi.

Mukhopadhyay, S.K. (1992), Distribution of Interarrival Time for Dry and Wet Years, *Environment and Ecology*, 10 (2), 424-426.

Ochi, K.M. (1989), *Applied Probability and Stochastic Processes in Engineering and Physical Sciences*, John Wiley & Sons, NY, USA.

Singh, G. (1973), Long Dry Spells of Short Return Periods during Southern Monsoon along the Konkan Coast, *Indian Met. Geophys.*, 24(1), 35-36.

Walker, G. (1919), Memoirs of the Indian Meteorological Department.

18

# Customer Satisfaction : An Overview from the Banking Sector in India

*R. K. Das, A. K. Das and S. R. Pal*

## 1. Introduction

On the verge of a new millennium we are facing the world in a new dimension. In this new era the traditional understandings are taking new shapes, particularly with the introduction of "Economic Globalization". A few years back, we were not much familiar with such words like liberalization, customer satisfaction, consumer forum, etc. But with the upliftment of our maturity level in open market policy and advancement of new technologies, we come to know the betterment of daily living. The buying of goods and services is not only for fulfilling the day-to-day wants, but to utilize every unit of the hard-earned money, So not only the producers will make sure about their profit by selling their products of services, but also the customers have full rights to get proper satisfaction from buying those products. Hence, it is gaining prime importance to measure the level of customer satisfaction from the consumption of both goods and services offered by the private sector. However, the measurement of the level of customer satisfaction from the consumption of any product, say an apple, is relatively easier than that of services like banking service, railways service, telecommunication service, etc. In India not much work has been carried out in this regard. In this paper, an attempt has been made to measure the level of customer satisfaction from banking services offered by both public as well as private sectors.

## 2. Materials and Methods

Attempts have been made to measure the extent of smooth and quick services towards the general people (customers here) from different banks. Customers are said to be satisfied when they are served quickly and smoothly from different banks for different purposes. Here we define the customer satisfaction parameter (U%) through the time consumed by the sectors to provide any service towards the general customers. It is not easy to incorporate the mental status of customers while measuring customer satisfaction, because the relationship between the times spent for a particular service and feeling impatient, varies widely among the customers. We have defined the mean service time for a specific service as the actual time taken by that bank to provide it to the customers and median time as the ideal time taken to serve 50 per cent of the gathered customers desiring that service. We then take the difference between the mean time and the median time and to standardize the difference to make the customer satisfaction parameter unit free, we divide that difference by the standard deviation of time consumed by that bank to provide that service. We then multiply it to standardize the parameter of customer satisfaction by 100 to make the customer satisfaction measurement scale into 100 equal divisions. The range of customer satisfaction parameter varies from + 100 to --100 i.e., $-100 \leq U\% \leq -100$, where we define,

$$U\% = \frac{Mean - Median}{S.D.} \times 100.$$

We are defining that service of a bank as providing satisfactory customer service when the mean service time of a service of a sector exceeding the median service time of that service of that bank. In other words, if 50 per cent of the gathered customers for a service are served within the mean service time, then we tend to define the service of that sector to be satisfactory. Hence, if U% is positive, then our conclusion is that customers are satisfied for that service. If U% equals zero (0), then we are calling that service of a bank as providing neither unsatisfactory nor satisfactory service, and if U% is negative, then the service of that bank is unsatisfactory. Hence, customers always want the U% of any service of any bank to be positive and equals to + 100. On the other hand, customers tend to avoid a service of any bank if its U% is negative and move towards -100. So, if a sector has provided the U% of a service as + 100, then the customers are defined to be extremely satisfied, and if it is –100, then customers are defined to be extremely unsatisfied.

Here U% parameter is well balanced in either way as its two hands measuring equal length of customer satisfaction. One for positive U and other for the negative one. It helps us to compare the relative performances of different sectors, as for example, +50 of U for a sector and –50 of U of a sector can be easily compared and we can easily say that the former sector performs doubly better than the later one as (–50 – (+50)}= –100 = 2 x (–50).

Efforts have also been made to fit the collected data for different services from different banks to Weibull distribution, a modified exponential distribution introduced by D.G. Kendal (1953). It is a multiple server queuing model, which describes the characteristics of arrival time distribution (of customer here), departure time distribution and also service time distribution. He used exponential distribution for service time distribution. The Weibull distribution has been introduced by Waloddi Weibull, a Swedish physicist, in 1939 and represented by the function :

$$f(x) = \alpha\beta . x^{\alpha-1} e^{-\beta . x^{\alpha}}, \alpha > 0, \beta > 0, 0 < x < \alpha,$$

where α and β are the two parameters and $x$ represents the service time. The mean of the function is given by :

$$E(x) = \int_0^\infty x_\wedge^f(x)dx = \frac{1}{\beta^{\frac{2}{\alpha}}}\Gamma\left(\frac{1}{\alpha}+1\right) \qquad \ldots (1.1)$$

and α can be solved through solving the ratio given by :

$$\text{var}(x) = \frac{1}{\beta^{\frac{2}{\alpha}}}\left[\Gamma\left(\frac{2}{\alpha}+1\right) - \left\{\Gamma\left(\frac{1}{\alpha}+1\right)\right\}^2\right] \qquad \ldots (1.2)$$

Thus

$$1+\frac{\text{var}(x)}{\{E^2(x)\}} = \frac{\Gamma\left(\frac{2}{\alpha}+1\right)}{\left\{\Gamma\left(\frac{1}{\alpha}+1\right)\right\}^2},$$

and β can solved by substituting the value of α either in equation (1.1) or in equation (1.2).

$$\int_0^{Me} f(x)dx = \frac{1}{2}$$

$$\text{or} \int_0^{Me} \alpha\beta^{a-1}.e^{-\beta x^a}dx = \frac{1}{2}$$

Let us put. $\beta x^\alpha = y$ or, $\alpha\beta x^{a-1}.dx = dy$

Hence, $x \to 0 \times y \to 0$, and $x \to Me \times y \to \beta Me^a$. Then

$$\int_0^{\beta.Me^\alpha} e^{-y}{}_{dy} = \frac{1}{2}$$

$$\text{or} \quad \left[-e^{-y}\right]_0^{\beta - Me^a} = \frac{1}{2}$$

$$\text{or} \quad 1 - e^{-\beta . Me^{\alpha}} = \frac{1}{2}$$

$$\text{or} \quad e^{-\beta . Me^{\alpha}} = \frac{1}{2}.$$

We know the formulae for determining the median of a continuous distribution as :

$$Me = \left(\frac{0.69314718}{\beta}\right)^{\frac{1}{\alpha}}.$$

Taking log on both sides we get

$$-\beta . Me^{a} . \log_e e = \log_e 0.5$$

$$\text{or} \quad -\beta . Me^{a} = -0.69314718.$$

Now, for any customer, the priority of a service of any sector is U% ≥ 0

$$\text{or} \quad \frac{Mean - Median}{S.D.} \times 100 \geq 0$$

$$\text{or} \quad Mean - Median \geq 0$$

$$\text{or} \quad \frac{1}{\beta^{\frac{1}{\beta}}} \Gamma\left(\frac{1}{\alpha} + 1\right) - \left(\frac{0.69314718}{\beta}\right)^{\frac{1}{\alpha}} \geq 0$$

$$\text{or} \quad \frac{1}{\beta^{\frac{1}{\alpha}}} \left[\frac{1}{\alpha} \Gamma\left(\frac{1}{\alpha}\right) - (0.69314718)^{\frac{1}{\alpha}}\right] \geq 0$$

$$\text{or}\ \frac{1}{\alpha}\Gamma(\frac{1}{\alpha}) - (0.69314718)^{\frac{1}{\alpha}} \geq 0 \ [\text{since } \beta > 0, \alpha > 0 \times \beta^{\frac{1}{\alpha}} > 0]$$

$$\text{or}\ \left(\frac{1}{\alpha}\Gamma(\frac{1}{\alpha})\right)^{\alpha} \geq 0.69314718.$$

Hence, from the above expression in terms of $\alpha$ we can say that it is the a parameter that actually directs U% to be positive or negative. That is why we are calling a parameter as the efficiency parameter, whereas, the other parameter $\beta$ has no ultimate effect on U%. This is all the theoretical explanation of U%.

After fitting the collected data to the Weibull function, a $\chi^2$ -tests has been carried out to test the goodness of this fit.. But some of our collected responses of some services of certain banks are too small to proceed for a parametric test (chi-square test) for goodness-of-fit. In other words, the collected frequencies of certain services of some banks consist of either null frequency or the frequencies which are less than five. We have then merged some of the classes whose respective frequencies are less than five, so that the cumulative frequency of those classes comes out to be five or greater than five. We have presumed that the distribution of any service of any sector should follow a theoretical distribution in the form of Weibull distribution. There are two parameters in the Weibull distribution ($\alpha$ and $\beta$). We all know that the degrees of freedom of $\chi^2$ for testing the goodness-of-fit is (n-1-k) where n is the number of classes after merging and k is the number of parameters to be estimated (here it is two). In other words, the degrees of freedom for testing the goodness-of-fit to the Weibull distribution is (n-1-2) or simply (n-3). But after the merge operation to the classes to make the class frequency at least five, the ultimate degrees of freedom of $\chi^2$ turned out to be 0 or even negative, which is meaningless. So to get rid of this problem we have followed the technique laid down by Kolmogorov (1933) and Smirnov (1939), to test the goodness of-fit of one sample to a specified distribution (here it is Weibull distribution), popularly known as K-S non-parametric test for the goodness-of-fit for one sample to a specified distribution.

The Kolmogorov-Smirnov one sample test is a test of goodness-of-fit. That is, it is concerned with the degrees of agreement between the distribution of a set of sample values (observed scores) and some specified theoretical distribution. It determines whether the scores in the sample can reasonably be thought to have come from a population having the theoretical distribution.

The one sample Kolmogorov-Smirnov test procedure compares the observed cumulative distribution function for a variable with a specified theoretical distribution. The Kolmogorov-Smirnov $D_n$ is computed from the largest difference (in absolute value) between the observed $S_n(x)$ and theoretical $F_0(x)$ cumulative distribution functions. This goodness-of-fit test whether the observations could reasonably have come from the specified distribution. Mathematically, Kolmogorov-Smirnov established that

$$D_n = max \mid F_n(x) - S_n(x) \mid,$$ where $n$ is the total number of scores.

The computed $D_n$ then compared with the tabular values of $D_n$ both at the 5% and 1 % levels of significance. If the computed $D_n$ is less than the tabulated value of $D_n$, then the null hypothesis is not rejected, otherwise $H_0$ is rejected, where

$H_0$: the sample has come from the specified theoretical distribution, against

$H_1$: the sample has not come from the specified theoretical distribution

We have seen that the probability density function of a continuous variable following Weibull distribution is

$$f(x) = \alpha\beta x^{\alpha-1}\bar{e}^{\beta x^{\alpha}}$$

and if $\alpha = 1$, then

$$f(x) = \beta e^{-\beta x},$$

which is nothing but the exponential distribution.

Now the coefficient of skewness. $\gamma_1 = 1/\beta^2 > 0$ for all values of $\beta$. In other words, exponential distribution is positively skewed where median is less than the mean. Similarly, Weibull distribution also is a positively skewed distribution whose median is less than the mean of the Weibull distribution. Hence, we can easily say that if the frequency distribution of any service of any bank has followed the Weibull distribution, then U% of that service must be positive, or in other words, the customers are satisfied if the frequency distribution of any service of any bank has been fitted good into the Weibull distribution, otherwise not.

We know that to construct a sampling frame to collect information regarding the population under study, the foremost condition is that the population must exist and whether it is homogeneous or heterogeneous. Here in case of banks, the population does exist but it is not known whether they are homogeneous or heterogeneous. Hence for our study it is not possible to construct a sampling frame to collect necessary information. That is why we have selected randomly any bank at any region and study them.

To collect data we have chosen to Nationalized banks (State Bank of India and United Bank of India) and two foreign Banks (Hong Kong Shanghai Banking Corporation and Australia-New Zealand Grindleys) and specifically two Nationalized banks in semi-urban areas (Kalyani) with a low density of population and two foreign banks in metro city (Kolkata) with a high density of population.

We have collected 134 customer responses from State Bank of India, 184 customer responses from United Bank of India, 133 customer responses from Hong Kong Shanghai Banking Corporation and 148 customer responses from Australia-New Zealand Grindleys. Altogether 599 customer responses were collected from the banking sector. This survey has been conducted through well-constructed questionnaire for separate sector separately for over five months.

After collecting the raw data from the field we have constructed a frequency distribution following a class interval of five minutes for initial thirty minutes and twenty minutes for the remaining time. From the frequency distribution we have calculated the mean, the median and the standard deviation (σ).

## 3. Results and Discussion

### *Performance of customer satisfaction of different banks for the fixed deposit service*

We start with the case of the Fixed Deposit service under the Savings category provided by the banks (State Bank of India, United Bank of India, Hong-Kong Shanghai Banking Corporations and Australia New Zealand Grindleys). According to the proposition, if the Customer Satisfaction parameter U% is positive, then it will fit good to the Weibull distribution. In other words, the customers are said to be satisfied if the U% parameter is positive and automatically the fitting will be good to the Weibull distribution. Here, we have seen from the Table 18.1 that the State Bank of India, United Bank of India and Australia New Zealand Grindleys have fulfilled the criteria of possessing positive U% and good bit to the Weibull distribution. To test the goodness-of-fit to the Weibull distribution we have followed the Kolmogorov-Smirnov non-parametric test. Let us see the Table 18.1 for K-S non-parametric tests of the said banks respectively :

**Table 18.1 :** Performance of customer satisfaction of different sectors for the Fixed Deposit Service

| *Bank* | $D_n$ | $D_{crit}$ | | *U%* |
|---|---|---|---|---|
| | | *5%* | *1%* | |
| State Bank of India | 0.099427914 | 0.563 | 0.669 | 19.42571725 |
| United Bank of India | 0.122990034 | 0.624 | 0.734 | 15.89805089 |
| Hong-Kong Shanghai Banking Corporations | 0.035473213 | 0.349 | 0.418 | -2.506145 |
| Australia New-Zealand Grindleys | 0.02864219 | 0.264 | 0.317 | 10.947843 |

Table 18.2 presents the necessary calculation for K.S. test for S.B.I data.

**Table 18.2.** Fitting to the Weibull distribution of the Fixed Deposit Service Response of the State Bank of India

| *U.C.B.(x) in minutes* | *Expected Responses* | *Observed Responses* | $F_0(x)$ | $S_n(x)$ | $\mid F_0(x) - S_n(x) \mid$ |
|---|---|---|---|---|---|
| 5.5 | 0.015418214 | 0 | 0.003083643 | 0 | 0.003083643 |
| 10.5 | 0.105377085 | 0 | 0.02415906 | 0 | 0.02415906 |
| 15.5 | 0.286837871 | 0 | 0.081526634 | 0 | 0.081526634 |
| 20.5 | 0.531570812 | 1 | 0.187840796 | 0.2 | 0.012159204 |
| 25.5 | 0.770018688 | 1 | 0.341844534 | 0.4 | 0.058155466 |
| 30.5 | 0.909687409 | 1 | 0.523782016 | 0.6 | 0.076217984 |
| 35.5 | 0.883950353 | 1 | 0.700572086 | 0.8 | 0.099427914 |
| 60.5 | 1.49714 | 1 | 1 | 1 | 0 |
| Total | 5 | 5 | | | |

Table 18.1 tells us that the $D_n$ for one-sample K-S non-parametric test is 0.099427914, which is less than the tabulated $D_{5,0.05}$ (0.563) and $D_{5,0.01}$ (0.669) for two-tailed test. Hence, the null hypothesis ($H_0$) is not rejected implying that the collected sample from the State Bank of India for the Fixed Deposit service has come from the Weibull distribution. On the other hand the U% is 19.42571725, we can definitely conclude that the fixed deposit service of the State Bank of India has followed the criteria of customer satisfaction in positive

direction. In other words, the State Bank of India has provided satisfactory customer service for the fixed deposit purpose.

In the same way, the results for the United Bank of India, Hong Kong Shanghai Banking Corporations and the Australia - New Zealand Grindleys tell us that (a) in case of United Bank of India the $D_n$ is 0.122990034, against the tabulated $D_{4,0.05}$ (0.624) and $D_{4,0.01}$ (0.734) for two-tailed test and on the other hand the U% is 15.89805089. Hence, the null hypothesis ($H_0$) is not rejected implying that the sample from United Bank of India for the fixed deposit service is following Weibull distribution and its U% is positive, we can conclude that the fixed deposit service of the United Bank of India has followed the criteria of customer satisfaction in positive direction. In other words, the United Bank of India has provided satisfactory customer service for the fixed deposit purpose; (b) in case of Hong-Kong Shanghai Banking Corporations the $D_n$ is 0.035473213, against the tabulated $D_{15,0.05}$ (0.349) and $D_{15,0.01}$ (0.418) for two-tailed test but the U% is -2.506145. This is not justified to conclude that the Hong-Kong Shanghai Banking Corporations has proved satisfactory or unsatisfactory customer service; (c) for the Australia New-Zealand Grindleys the $D_n$ is 0.02864219 against the tabulated $D_{25,0.05}$ and $D_{25,0.01}$ (0.317) and the U% is 10.94784313, which tells us that the Australia New-Zealand Grindleys bank has provided satisfactory customer service.

### *Performance of customer satisfaction of different sectors for the savings deposit service*

The service regarding the Savings Deposit under the Savings category provided by the banks (State Bank of India, United Bank of India, Hong-Kong Shanghai Banking Corporations and Australia New-Zealand Grindleys) successfully demonstrate the proposition of possessing positive U% and good fit to the Weibull distribution following K-S non-parametric test. In other words, both the sectors have served satisfied customer service for the Savings Deposit under the Savings category. Let us show the result in the Table 18.3.

**Table 18.3 :** Performance of Customer Satisfaction of Different Sectors for the Savings Deposit Service

| *Bank* | $D_n$ | $D_{crit}$ | | *U%* |
|---|---|---|---|---|
| | | *5%* | *1%* | |
| State Bank of India | 0.060457322 | 0.198 | 0.237 | 18.89650954 |
| United Bank of India | 0.028949681 | 0.190 | 0.228• | 14.02975002 |
| Hong-Kong Shanghai Banking Corporations | 0.007340958 | 0.217 | 0.261 | 17.47495781 |
| Australia New-Zealand Grindleys | 0.018310527 | 0.202 | 0.242 | 23.17808294 |

Table 18.3 demonstrates the fact that all the banks have provided satisfactory customer service and fitted the Weibull distribution fits well, both at the 5% and 1 % levels of significance with 5 degrees of freedom except the Hong-Kong Shanghai Banking Corporations. This is because of the fact that the class frequency of the Hong-Kong Shanghai Banking Corporations is zero, so they are merged together which results in minimizing the class intervals. As a result of this the degrees of freedom for testing the goodness of fit has come down to three, but till then the fitting to the Weibull distribution is good. Out of all the sectors Australia New-Zealand Grindleys has come out to be the best among all the banks to provide customer satisfaction regarding the Savings Deposit service under the Savings category.

*Performance of customer satisfaction of different sectors for the withdrawal of cash service*

Regarding the service of the withdrawal of cash from the State Bank of India, United Bank of India; Hong-Kong Shanghai Banking Corporations, Australia New Zealand

Grindleys and the Post Office where the customers' interest lie heavily as the customers draw their hard-earned money on emergency has thrown a contradictory result towards measuring customer satisfaction. Because, most of the sectors except the Post Office have contradicted with the criteria of positive U% and good fit to the Weibull distribution. This may be due to some extraordinary characters which is unfortunately unknown to us. Let us see the fact in Table 18.4.

**Table 18.4 :** Performance of Customer Satisfaction of Different Sectors for the Withdrawal of Cash Service

| *Bank* | $D_n$ | $D_{crit}$ | | *U%* |
|---|---|---|---|---|
| | | *5%* | *1%* | |
| State Bank of India | 0.861743483 | 0.205 | 0.245 | 41.57236879 |
| United Bank of India | 0.922464368 | 0.188 | 0.226 | 15.36145585 |
| Hong-Kong Shanghai Banking Corporations | 0.055071439 | 0.198 | 0.237 | -4.87904394 |
| Australia New-Zealand Grindleys | 0.08831612 | 0.196 | 0.235 | -12.1314920 |

Table 18.4 shows a very disappointing picture of the customer satisfaction of the withdrawal of cash service. None of the above sectors have served satisfactory service to their customers. In some cases, the U% is positive but their fitting to the Weibull distribution following K-S non-parametric test is not good, whereas, in some cases the U% is negative although their fitting to the Weibull distribution following K-S non-parametric test is good. Hence, no definite conclusion following our proposal of Customer Satisfaction can be drawn from the withdrawal of cash service. The confusing behaviour to the overall picture for the withdrawal of cash category may be due to the fact that the collected sample from the different sectors falls short of frequencies (as 44 responses are recorded for the State Bank of India, 52 responses are recorded for the United Bank of India; 47

responses are recorded for the Hong Kong Shanghai Banking Corporations, 48 responses are recorded for the Australia New Zealand Grindleys and 71 responses are recorded for the Post Office) that causes the ill demonstration of the fact. But in general all the sectors have failed to deliver customer satisfaction regarding the most desired service for any customer from any financial institution, the withdrawal of cash. In other words, the sectors are not taking care of the financial transaction from the sectors towards the customers. This in turn may create problem for the financial institutions in future as the customers feel that they are neglected in this regard.

*Performance of customer satisfaction of different sectors for the current account service*

For the service of Current Account, all the banks except the United Bank of India follow our proposition of positive U% and good fit to the Weibull distribution which is shown in the Table 18.5.

**Table 18.5 :** Performance of customer satisfaction of different sectors for the current Account Service

| *Bank* | $D_n$ | $D_{crit}$ | | *U%* |
|---|---|---|---|---|
| | | *5%* | *1%* | |
| State Bank of India | 0.053729296 | 0.430 | 0.513 | 25.25907 |
| United Bank of India | 0.280633812 | 0.624 | 0.734 | 35.33571 |
| Hong-Kong Shanghai Banking Corporations | 0.031490979 | 0.519 | 0.617 | 11.27752 |
| Australia New Zealand Grindleys | 0.04204997 | 0.430 | 0.513 | -12.1315 |

Table 18.5 demonstrates that the State Bank of India, United Bank of India, Hong-Kong Shanghai Banking Corporations and the combined efforts of all the Banks have put up very conscious efforts about the Current Account

service towards their customers as their U% is positive and they possess a good fit to the Weibull distribution following K-S non-parametric test. Hence, the customers have been satisfied for the service of Current Account from the sectors. Only, the Australia New Zealand Grindleys may be due to some exogenous characters does not obey our proposition of Customer Satisfaction for the service of Current Account.

*Performance of customer satisfaction of different sectors for the visit to the manager regarding loan (either in the form of cash or letter of credit) service*

Our findings regarding the service provision of a visit to the manager regarding loan (either in the form of cash or letter of credit) are demonstrated through Table 18.6.

**Table 18.6 :** Performance of Customer Satisfaction of Different Sectors for the Visit to the Manager regarding Loan (either in the form of cash or letter of credit)

| *Bank* | $D_n$ | $D_{crit}$ | | *U%* |
|---|---|---|---|---|
| | | *5%* | *1%* | |
| State Bank or India | 0.146243102 | 0.708 | 0.829 | 33.74317937 |
| United Bank or India | 0.096849758 | 0.361 | 0.432 | 25.71314784 |
| Hong-Kong Shanghai Banking Corporations | 0.021983887 | 0.483 | 0.576 | 9.628954139 |
| Australia New Zealand Grindleys | 0.04921905 | 0.519 | 0.617 | 13.81838146 |

Table 18.6 demonstrates that the State Bank of India, the United Bank of India, the Hong-Kong Shanghai Banking Corporations, the Australia New Zealand Grindleys and also the combined efforts of all the banks meet the Customer Satisfaction criteria in positive direction for the discussion with the manager regarding loan (either in the form of cash or letter of credit), i.e., the U% of the State Bank of India, the United Bank of India, the Hong-Kong Shanghai Banking Corporations, the Australia New-Zealand Grindleys and the

combined efforts of all the banks are all positive and their fitting to the Weibull distribution are good following K-S non-parametric test both at the 5% and 1 % levels. Hence, the State Bank of India, the United Bank of India, the Hong-Kong Shanghai Banking Corporations, the Australia New Zealand Grindleys and the combined efforts of all the Banks do care about discussion with the manager regarding loan (either in the form of cash or letter of credit).

*Performance of customer satisfaction of different sectors for the visit to the manager regarding any specific problem or discussion with manager-in-charge in lieu of manager service*

For the service provision of customer grievances over the overall service of the banks or regarding any specific problem of customer towards the banker, we have collected responses together under a single column of discussion with the manager regarding any specific problem or discussion with manager-in-charge in lieu of manager. Let us present the result through Table 18.7.

**Table 18.7 :** Performance of Customer Satisfaction of different sectors for the Visit to the Manager regarding any Specific problem or discussion with Manager-in-charge in lieu of manager service

| *Bank* | $D_n$ | $D_{crit}$ | | *U%* |
|---|---|---|---|---|
| | | *5%* | *1%* | |
| State Bank of India | J.098687246 | 0.318 | 0.381 | 18.22538 |
| United Bank of India | 0.072996943 | 0.291 | 0.350 | 28.82961 |
| Hong-Kong Shanghai Banking Corporations | 0.048360454 | 0.301 | 0.361 | 28.312795 |
| Australia New-Zealand Grindleys | 0.026176152 | 0.338 | 0.404 | -0.967551 |

Table 18.7 demonstrates that the State Bank of India, the United Bank of India, the Hong-Kong Shanghai Banking Corporations and also the combined efforts of all the banks

meet the Customer Satisfaction criteria in positive direction for the discussion with the manager regarding any specific problem or discussion with Manager-in-charge in lieu of manager, i.e., the U% of the State Bank of India, the United Bank of India, the Hong-Kong Shanghai Banking Corporations and the combined efforts of all the banks are all positive and also their fit to the Weibull distribution are good following K-S non-parametric test. Hence, the State Bank of India, the United Bank of India, the Hong-Kong Shanghai Banking Corporations and the combined efforts of all the banks do care about discussion with the manager regarding any specific problem or discussion with manager-in-charge in lieu of manager. But the performance of the Australia New Zealand Grindleys regarding Customer Satisfaction for the discussion with the manager regarding any specific problem or Discussion with Manager-in-charge in lieu of manager is contradictory may be due to the presence of some exogenous factors, because on the one hand its U% is negative but on the other hand its fitting to the Weibull distribution following K-S non-parametric test is good. Hence, no definite conclusion can be drawn regarding Customer Satisfaction of the Australia New Zealand Grindleys in this regard.

*Performance of customer satisfaction of different sectors for the draft making service*

Lastly, we also collect responses regarding draft making which is an important service from customers' point of view as it involves special security to the customers while transacting cash. In this context we shall confine our discussion within the Nationalized banks only, as the foreign banks do not offer this sort of service towards their customers. Let us see the result regarding this performance of Customer Satisfaction through Table 18.8.

**Table 18.8 :** Performance of Customer Satisfaction of Different Sectors for the Draft Making Service

| *Bank* | $D_n$ | $D_{crit}$ | | *U%* |
|---|---|---|---|---|
| | | *5%* | *1%* | |
| State Bank of India | 0.0453354216 | 0.430 | 0.513 | 73.77009274 |
| United Bank of India | 0.161579282 | 0.218 | 0.261 | 11.58582917 |
| All banks together | 0.106389673 | 0.196 | 0.235 | 17.33473 |

Table 18.8 demonstrates that the State Bank of India, the United Bank of India and also the combined efforts of all Nationalized banks meet the Customer Satisfaction criteria in positive direction for the draft making service, i.e., the U% of the State Bank of India, the United Bank of India and the combined efforts of all Nationalized banks are all positive and also their fittings to the Weibull distribution are good following K-S non-parametric test. Hence, the State Bank of India, the United Bank of India and the combined efforts of all Nationalized banks do care about draft making service.

*Performance of customer satisfaction of all the banks considering all sorts of services*

Let us see the performances of different banks considering all sorts of service they provide. We start our discussion in this regard with the State Bank of India. We have considered seven services provided by the banks such as : I. Fixed Deposit under the Savings category; 2. Savings Deposit under the Savings category; 3. Withdrawal of Cash; 4. Current Account; 5. Discussion with the manager regarding loan (either in the from of cash or letter of credit); 6. Discussion with the manager regarding specific problem; and 7. Draft-making (which is provided by the two Nationalized banks only). Let us sum up the performances of different services of the State Bank of India in Table 18.9.

**Table 18.9 :** Performance of Customer Satisfaction of all the banks considering all sorts of services

| *Bank* | $D_n$ | $D_{crit}$ | | *U%* |
|---|---|---|---|---|
| | | *5%* | *1%* | |
| State Bank of India | 0.118731128 | 0.118 | 0.140 | 37.24511605 |
| United Bank of India | 0.076696952 | 0.100 | 0.120 | 30.250451 |
| Hong-Kong Shanghai Banking Corporations | 0.02599772 | 0.117 | 0.141 | 10.46777408 |
| Australia New Zealand Grindleys | 0.035098997 | 0.111 | 0.133 | 7.523641356 |

Table 18.9 shows that the overall performance of all the banks except the State Bank of India towards Customer Satisfaction is satisfactory considering all sorts of services, as U% comes to be positive and their fitting to the Weibull distribution is good following K-S non-parametric test. But in case of the State Bank of India, the U% is positive but its fitting to the Weibull distribution is marginally rejected following K-S non-parametrics test, both at the 5% and 1% levels of significance. That means that following the concept of our test criteria of satisfactory service provider we can claim directly that the United Bank of India, Hong-Kong Shanghai Banking Corporations and Australia New Zealand Grindleys provide satisfactory service towards their customers out of which United Bank of India is the best followed by Hong Kong Shanghai Banking Corporations and Australia New Zealand Grindleys.

## 4. Summary and Conclusion

The concept of Customer Satisfaction has started in the United States of America during 1930s. The literal meaning of Customer Satisfaction simply is the level of utility that a consumer gains from the consumption of any goods or service available in the economy. The developed countries like the United States of America, the United Kingdom, Germany, France, Japan etc. have developed several techniques to improve the Customer Satisfaction from the producers' as well as from the consumers' point of view. But in India, a developing economy has not yet settled down to provide the best Customer Satisfaction to the consumers particularly, from the service sector's point of view. In the consumption of goods sector we already have several procedures to measure the Customer Satisfaction. For example, after viewing the electricity consumption of a refrigerator, how fast it makes ice, how much it costs, how it looks like, what is its capacity etc. one can compute the level of satisfaction obtainable from that refrigerator, or by examining the load capacity of a tyre, the speed limit of the vehicle to stop it smoothly using that tyre, the price of the tyre etc. one can compute the level of satisfaction from the tyre. But, unfortunately, in India there is still no such technique to measure the level of Customer Satisfaction from the services that the consumers consume. For example, there are no such defined measures to examine whether the banking or non-banking financial institutions provide the level of satisfaction what the customers desire, or there are no such defined procedures to measure the level of Customer Satisfaction from the services provided by the Railways, Airlines, Ships or by the Telecommunications. Even we do not have any such procedures to measure the level of Customer Satisfaction from the services offered by the hospitals, the nursing homes etc. We just have proposed a technique to measure directly the level of Customer Satisfaction from the different services provided by different sectors. In simple language, the objective of this study is to provide a tool using which one can measure the level of Customer Satisfaction from the service he or she is exercising.

Applying the proposed techniques of U% and the one-sample K-S non-parametric test for goodness-of-fit to the Weibull distribution, we conclude that if the U% is positive, the collected sample for a specific service from any relevant sector will follow the Weibull distribution. Hence, we can

**Table 18.10 :** Performance of Customer Satisfaction of Different Services offered by Different Banks

| *Service period* | *Service performance of the banks* | *Remarks* |
|---|---|---|
| Fixed deposit | State Bank of India | Satisfactory service |
| | United Bank of India | Satisfactory service |
| | Hong-Kong Shanghai Banking Corporations | Satisfactory service |
| | Australia New Zealand Grindleys | Unsatisfactory service |
| Savings Deposit | State Bank of India | Satisfactory service |
| | United Bank of India | Satisfactory service |
| | Hong Kong Shanghai Banking Corporations | Satisfactory service |
| | Australia New Zealand Grindleys | Satisfactory service |
| Withdrawal of Cash | State Bank of India | Unsatisfactory service |
| | United Bank of India | Unsatisfactory service |
| | Hong Kong Shanghai Banking Corporations | Unsatisfactory service |
| | Australia New Zealand Grindleys | Unsatisfactory service |
| Current Account | State Bank of India | Satisfactory service |
| | United Bank of India | Satisfactory service |
| | Hong-Kong Shanghai Banking Corporations | Satisfactory service |
| | Australia New-Zealand Grindleys | Unsatisfactory service |
| Visit to the Manager regarding loan | State Bank of India | Satisfactory service |
| | United Bank of India | Satisfactory service |
| | Hong-Kong Shanghai Banking Corporations | Satisfactory service |
| | Australia New Zealand Grindleys | Satisfactory service |
| Visit to the Manager regarding any specific problem | State Bank of India | Satisfactory service |
| | United Bank of India | Satisfactory service |
| | Hong Kong Shanghai Banking Corporations | Satisfactory service |
| | Australia New Zealand Grindleys | Unsatisfactory service |

say that the service provided by any sector is said to be satisfied if the U% is positive and simultaneously the sample follows the Weibull distribution. Let us present the results obtained from the analysis for different services from different sectors either individually or jointly (if the service is offered by more than one sector).

From the study of Customer Satisfaction from the public sectors in India we have seen that the services of the banking sectors as a whole is not conclusive using the proposed criteria as the U% is positive but the fitting to the Weibull distribution is not good. The rejection of the fitting to the Weibull distribution is marginal. So, neglecting that we can say that the performance of the banking sectors is more or less satisfactory.

## References

Anderson, E. W. (1994), Cross-Category Variation in Customer Satisfaction and Retention, *Journal of Marketing Letters*, 5(1), 19-30.

Das, R.K. (2002), A Quantitative Analysis of Customer Satisfaction in the Public Sectors in India, *Ph. D. Thesis.*

Fornell C. (1983), A Industrial Organisation and Customer Satisfaction Dissatisfaction, *Journal of Consumer Research*, 9, 403-412.

Kendal, M.G. and Stuwart, A. (1961), *Advanced Theory of Statistics*, Charles Griffin.

Ryan, M. J., Thomas, B. and Venkataram, R. (1995), Making Customer Satisfaction Measurement a Power Tool, *Journal of Marketing Research*, 7(3), 11-16.

19

# Use of ARIMA Model on Rice Production in Birbhum District

*Manas Kumar Sanyal and Swaraj Kumar Mukhopadhyay*

## 1. Introduction

The economy of Birbhum district is based on agriculture. Rice production of this district is 1157.4 thousand tonnes. This crop is covered about 75% contribution of the total cultivation in this district. Forecasting is one of the major aim in statistical analysis. Many sophisticated models are used to forecast the economic time series. One of them is Autoregressive Integrated Moving Average Model (ARIMA). The advantage of ARIMA modeling over the other univariate time series model is that besides displaying the intrinsic behaviour of the (generating process) variable it produces the smallest mean squared forecast error variance. But the limitation of ARIMA model is that it is only applicable to forecast the near future values, usually, within five years beyond the last case in the time series.

Box-Jenkins ARIMA model has been employed to predict the future values of rice production in Birbhum. Relevant references are Box and Jenkins (1976), Diebold (2001), Patterson (2000), Pankratz (1983), Gujarati (2003).

The aim of this study is to fit the ARIMA model in rice production of Birbhum and forecast production of rice up to year 2006-07. The ability to make such predictions successfully is obviously important for policy-makers for making their future plan. A model is said to be the best if it generates minimum errors or residuals along with a very limited number of parameters for proper analysis.

## 2. The Model

Here time series variables of rice production in Birbhum district of West Bengal have been used during the period 1937-38 to 2001-02.

Now, Y is modeled as follows :

$$Y_t = \mu + \theta_0 u_t + \theta_1 u_{t-1},$$

where $\mu$ is a constant and $\mu_t$'s are the white noise stochastic error term. So, Y at time t is equal to a constant plus moving average of the current and one or more previous disturbances and we say that, $Y_t$ follows a first-order moving average or MA(1) process. In our study we apply ARIMA (0, 1, 1) model, i.e., the time series has to be differenced once, d=1, and it is a purely MA(1) process.

There is a model building procedure, described by Box and Jenkins (1976) that allows us to construct the best possible model for the above-mentioned series. This procedure consists of three steps : identification, estimation, and diagnosis. These three stages have to be repeated until the model is best fitted.

### 2.1 Identification of the Appropriate Model

The first and most subjective step is the identification of the processes underlying the series. In this stage the three integers, namely, p, d, and q in the ARIMA (p,d,q) process generating the series must be determined. Here p is the order of autoregression, d is the degree of differencing, and q is the order of moving average involved.

To identify the process underlying a series, thorough

testing has to be done to determine whether the series is non-stationary. A stationary series has the same mean and variance throughout. But, it is well-known that most of the economic time series are non-stationary. The major tools in identification are the autocorrelation function (ACF), the partial autocorrelation function (PACF), and the resulting correlograms, which are simply the plots of ACFs and PACFs against the lag length.

The methods used in identifying, estimating, and diagnosing ARIMA models are quite descriptive. The general model (non-seasonal), is traditionally written as ARIMA (p, d, q),

In the first step appropriate values of d is obtained to make the time series stationary. The parameter d is simply the number of times the series has been differenced to make it stationary. Next, the values of p and q must be identified, the orders of autoregression and moving average terms. Finally, the autocorrelation function (ACF) and partial autocorrelation function (PACF) of the series identifies the model.

### 2.2 Estimation

To estimate the parameters p, d, and q iterative calculations are required. The model is fitted by the minimum values of Akaike Information Criterion (AIC), Schwartz Bayesian Criterion (SBC), Estimated Correlation Coefficient (ECC), log-likelihood (LL) etc. and finally the parsimony of the model parameters p, d and q is verified.

## 3. Diagnostic Checking of the Models

After choosing a particular ARIMA model, and having estimated its parameters, testing has to be done to see whether the chosen model fits the data reasonably well. The simplest test is to observe whether the residuals estimated from this model are white noise. By testing the ACF of the residuals, if we observe that there is no autocorrelation of the residuals, then the residuals are really white noise. If they are, we accept the model, otherwise we repeat the process from start.

Also, to confirm the correctness or predictive power of the estimated best fitted model for forecasting the future values, the first 59 years' data, i.e., production of rice for the year 1937 to 1995 are taken. These cases are refitted in the respective models. Then, the actual observed values of production of rice and predicted values of production as obtained from the refitted ARIMA model are compared.

Finally, mean absolute per cent errors or MAPE (Pankratz, 1983) is computed. The MAPE measures the percentage of error present in the forecasting model and is computed as :

$$\text{Mean Absolute Percentage Error } (MAPE) = \frac{100}{m} \sum \frac{|e_{n+l}|}{Y_{n+l}}.$$

## 4. Results and Discussion

The time series data on production of rice in Birbhum district have been examined for periods of 1937-38 to 2001-02, a total of 65 observations for each time series. At the identification stage, by plotting the observed time series variable under study, i.e., rice production in Birbhum district is found to be non-stationary with respect to its mean and variance. By employing successive differencing the time series variable is found to be stationary at first differencing and so differencing of order one is taken. In the next stage, for getting the best model for production of rice in Birbhum district, several combinations of p and q are taken while fixing the value of d at 1. From the estimation of different parameters, ARIMA (0,1,1) is selected as the best fitted model for rice because this model possesses lowest SBC and AIC values. The log-likelihood values are comparatively larger and the standard errors of estimation are smaller for this model. The MAPE is also found to be smallest of about 7.61 per cent. The minimum value of the MAPE measures the accuracy of the model.

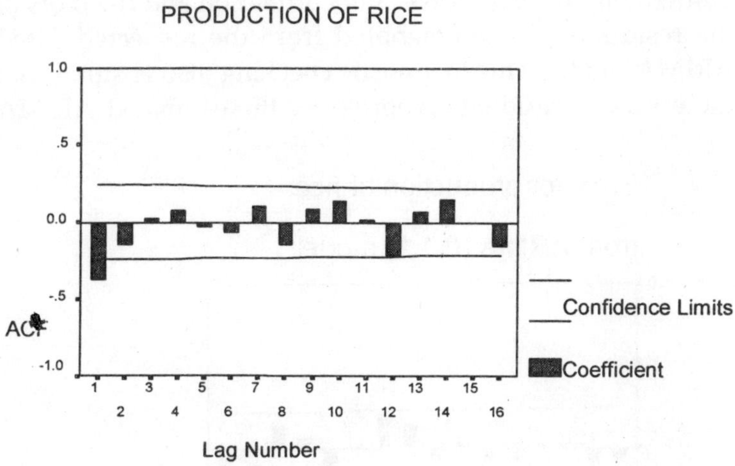


**Fig. 19.1:** Estimated ACF plot for Production of Rice in Birbhum District

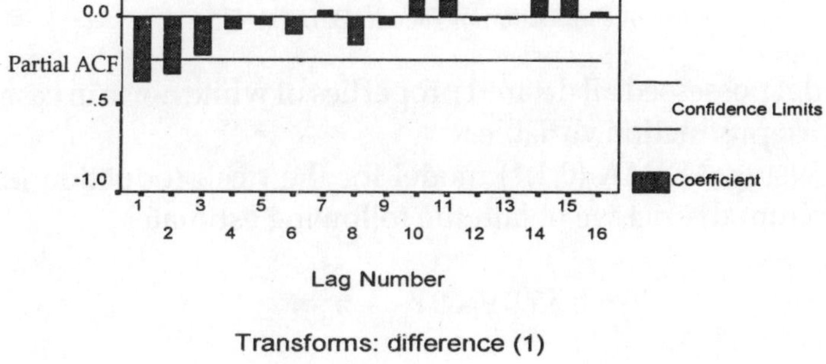


**Fig. 19.2 :** Estimated PACF plots for Prowduction of Rice in Birbhum District

The ACF and PACF plots for the differences in rice production are shown in Figures 19.1 and 19.2, respectively.

To diagnose the estimated ARIMA models with the residual series, ACF of the error series created by ARIMA is plotted. Figure 19.3 shows autocorrelation and the plots of the residuals of rice emanated from the respected fitted ARIMA model. This diagnostic checking also ensured that the errors or residuals reported by the estimated ARIMA

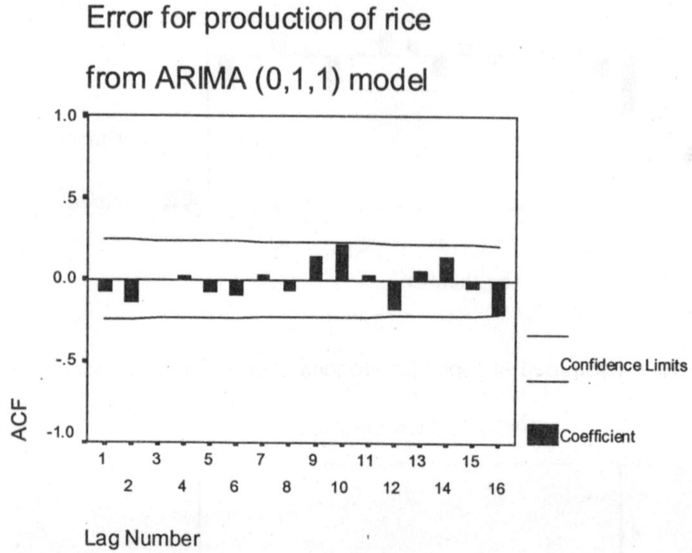


**Fig. 19.3 :** Estimated ACF plots of Residuals of the Fitted ARIMA (0,1,1) Model of Production of Rice in Birbhum District

model possessed all desired properties of white noise in case of rice production variable.

Using ARIMA (0,1,1) model for the rice production in Birbhum district we obtain the following estimates :

$$Y_t = 0.55979469Y_{t-1} + a_t,$$

where $a_t$ is the white noise term at time t.

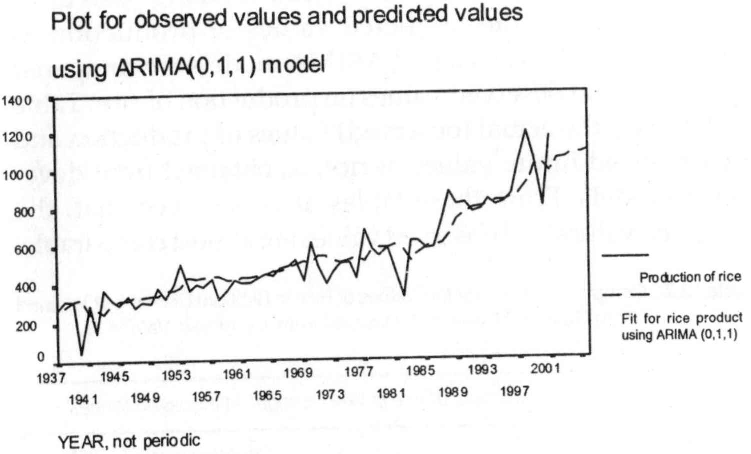


**Fig. 19.4 :** Plot for Observed Values for Rice Production and Predicted Values Obtained from ARIMA (0,1,1)

Table 19.1 displays the future values of the rice production in Birbhum district with their 95% probability range up to year 2007-2008.

**Table 19.1 :** Predicted Future Values of Rice in Birbhum District in West Bengal using ARIMA(0,1,1)-model

| *Years* | *Rice Production in Birbhum District in thousand tonnes using ARIMA (0,1,1) Model* | | |
|---|---|---|---|
| | *Predicted value* | *Confidence Interval = 95%* | |
| | | *Lower confidence limit* | *Upper confidence limit* |
| 2002-03 | 1039.28321 | 840.23684 | 1238.32959 |
| 2003-04 | 1051.62307 | 843.74373 | 1259.50240 |
| 2004-05 | 1063.96292 | 847.36314 | 1280.56270 |
| 2005-06 | 1076.30277 | 851.08197 | 1301.52357 |
| 2006-07 | 1088.64262 | 854.88923 | 1322.39601 |
| 2007-08 | 1100.98247 | 858.77558 | 1343.18937 |

Now, also to test correctness of the estimated models, the models are refitted for getting predicted values for the years 1937 to 1996, i.e., 60 years instead of taking cases up to year 2001. Then, the predicted values of production as obtained from the developed ARIMA models are compared with the actual observed values on production of rice. Table 19.2 depicts the actual (observed) values of production and the predicted future values of rice, as obtained from developed models. From these tables, it is observed that, the predicted values and observed values are almost comparable.

**Table 19.2 :** Comparison of Predicted Values of Rice in Birbhum District as Obtained from Refitted ARIMA (0,1,1) model with Observed Values

| *Years* | *Production of rice in West Bengal in thousand tonnes* | | | |
|---|---|---|---|---|
| | | | *Confidence level = 95%* | |
| | *Actual value* | *Predicted value* | *Lower confidence limit* | *Upper confidence limit* |
| 1997-98 | 1020.10 | 849.00389 | 677.14035 | 1020.86742 |
| 1998-99 | 1187.80 | 859.13942 | 680.33488 | 1037.94396 |
| 1999-2000 | 1042.50 | 869.27495 | 683.57743 | 1054.97246 |
| 2000-01 | 796.50 | 879.41048 | 686.86285 | 1071.95810 |
| 2001-02 | 1157.40 | 889.54600 | 690.18673 | 1088.90528 |

## References


Box, G. E. and Jenkins, G. M. (1976), *Time Series Analysis : Forecasting and Control*, Holden-Day, San Francisco, USA.

Diebold, F. (2001), *Elements of Forecasting*, South Western Publishing, USA.

Government of West Bengal (1997-98), *Statistical Abstract*, Bureau of Applied Economics and Statistics, Calcutta.

Government of West Bengal (2001-02), *Economic Review*, State Planning Board, Calcutta.

Gujarati, D. N. (1995), *Basic Econometrics*, 3rd Ed., McGraw-Hill International Edition.

Pankratz A. (1983), *Forecasting with Univariate Box-Jenkins Models: Concept and Cases*, John Wiley & Sons, New York.

Patterson, K. (2000), *An Introduction to Applied Econometrics : A Time Series Approach*, St. Martin's Press, New York.

# List of Contributors

1. **Agrawal Ranjana**
Indian Agricultural Statistics Research Institute,
Library Avenue, Pusa
New Delhi 110012
India

2. **Amdekar S. J.**
1/43 A-2 Pantnagar
District: U S Nagar (Uttaranchal)
PIN 263145
India

3. **Basu A. K.**
Department of Statistics
University of Calcutta
35 Ballygunge Circular Road
Calcutta 700019
India

4. **Bhattacharya Debasis**
Visva-Bharati University
Santiniketan
West Bengal
PIN 731236
India

5. **Bhattacharya Sankha**
Department of Statistics
Narendrapur Ramakrishna Mission Degree College
Narendrapur
Calcutta 700103
India

6. **Bhattacharyya Banjui**
Department of Agricultural Statistics
Bidhan Chandra Krishi Viswavidyalaya
Mohanpur, Nadia
West Bengal
PIN 741252
India

7. **Chakraborty Sourav**
National Sample Survey Organization
Mahalanobis Bhavan
164 G.L. Thakur Road
Calcutta 700108
India

8. **Chattopadhyay Asis Kumar**
Department of Statistics
University of Calcutta
35 Ballygunge Circular Road
Calcutta 700019
India

9. **Chaubey Yogendra P**
Concordia University
1455 de Maisonneuve Blvd. West
Montreal, Quebec
Canada H3G1M8

10. **Das A. K.**
Department of Agricultural Statistics
Bidhan Chandra Krishi Viswavidyalaya
Mohanpur, Nadia
West Bengal
PIN 741252
India

11. **Das R. K.**
Department of Agricultural Statistics

Bidhan Chandra Krishi Viswavidyalaya
Mohanpur, Nadia
West Bengal
PIN 741252
India

12. **Dwivedi S. N.**
Department of Biostatistics
All India Institute of Medical Sciences
Ansari Nagar
New Delhi 110029
India

13. **Ghosh Kunal**
Institute of Agricultural Science
University of Calcutta
35 Ballygunge Circular Road
Calcutta 700019
India

14. **Kole P. C.**
Visva-Bharati University
Santiniketan
West Bengal
PIN 731236
India

15. **Madhura B. P.**
Biomedical Data Sciences India
Glaxo SmithKline Pharmaceuticals
Bangalore 560068
India

16. **Majumdar N. R.**
Department of Agricultural Statistics
Uttar Banga Krishi Viswavidyalaya

Pundibari, Cooch Behar
West Bengal
PIN 736165
India

17. **Mukhopadhyay Swaraj K.**
Department of Agricultural Statistics
Bidhan Chandra Krishi Viswavidyalaya
Mohanpur, Nadia
West Bengal
PIN 741252
India

18. **Murty M. Vivekananda**
Department of Statistics and
Population Research Centre
Andhra University
Visakhapatnam
Andhra Pradesh
PIN 530003
India

19. **Pal S. R.**
Department of Agricultural Statistics
Bidhan Chandra Krishi Viswavidyalaya
Mohanpur, Nadia
West Bengal
PIN 741252
India

20. **Rao K. Srinivasa**
Department of Statistics
Andhra University
Visakhapatnam
Andhra Pradesh
PIN 530003
India

21. **Roychowdhury Soma**
Indian Institute of Social Welfare and Business Management, Management House,
College Square West,
Calcutta 700073
India

22. **Sanyal Manas Kumar**
Department of Business Administration
University of Kalyani, Nadia
West Bengal
PIN 741234
India

23. **Sen Roy S.**
Department of Statistics
University of Calcutta
35 Ballygunge Circular Road
Calcutta 700019
India

24. **Sundaram K. R.**
Department of Biostatistics
Amrita Institute of Medical Sciences and Research
Amrita Lane, Kochi
Kerala,
PIN 682041
India

25. **Vaman H. J.**
Department of Statistics
Central University of Rajasthan
R.K. Patni Govt. College Campus
Kashiangarh, Rajasthan
PIN 305802
India

**26. Varadachari Chandrika**
Raman Centre for Applied and
Interdisciplinary Sciences
16A Jheel Road
Calcutta 700075
India

# Index